Spon's Budget Estimating Handbook

Spon's Budget Estimating Handbook

Edited by
SPAIN AND PARTNERS
Consulting Quantity Surveyors

First Edition

E. & F.N. SPON

An imprint of Chapman and Hall
London · New York · Tokyo · Melbourne · Madras

UK	Chapman and Hall, 2–6 Boundary Row, London SE1 8HN
USA	Van Nostrand Reinhold, 115 5th Avenue, New York NY10003
Japan	Chapman and Hall Japan, Thomson Publishing Japan, Hirakawacho Nemoto Building, 7F, 1–7–11 Hirakawa-cho, Chiyoda-ku, Tokyo 102
Australia	Chapman and Hall Australia, Thomas Nelson Australia, 480 La Trobe Street, PO Box 4725, Melbourne 3000
India	Chapman and Hall India, R. Seshadri, 32 Second Main Road, CIT East, Madras 600 035

First edition 1990
© 1990 E. & F.N. Spon

Printed in Great Britain by
T. J. Press (Padstow) Ltd, Padstow, Cornwall

ISBN 0 419 14780 2   0 442 31176 1 (USA)

All rights reserved. No part of this publication may be reproduced or transmitted, in any form or by any means, electronic, mechanical, photocopying, recording or otherwise, or stored in any retrieval system of any nature, without the written permission of the copyright holder and the publisher, application for which shall be made to the publisher.

British Library Cataloguing in Publication Data
Spon's Budget estimating handbook.
  1. Construction industries. Costing
  I. Spain and Partners
  624.0681

ISBN 0 419 14780 2

Library of Congress Cataloging-in-Publication Data
Spon's budget estimating handbook/edited by Spain and Partners.—
—1st ed.
  p. cm.
  Includes index.
  ISBN 0 442 31176 1
  1. Building—Estimates. I. Spain and Partners. II. Title: Budget estimating handbook.
TH435.S7253 1990
692'.5—dc20   90–10005
                                                       CIP

Contents

Foreword by Martin Barnes	vii
Preface	viii
Introduction	x

PART ONE   BUILDING WORK
1	Square metre prices	3
2	Elemental costs	7
3	Composite rates	27

PART TWO   CIVIL ENGINEERING WORK
4	Principal rates	47
5	Composite rates	89
6	Project costs	97

PART THREE   MECHANICAL AND ELECTRICAL WORK
7	Square metre prices	119
8	Principal rates	129

PART FOUR   RECLAMATION, LANDSCAPING AND ENVIRONMENTAL WORK
9	Principal rates	153
10	Composite rates	161

PART FIVE   REFURBISHMENT
11	Composite rates	171

PART SIX   GENERAL DATA
12	Life cycle costing	177
13	The development process	183

14	Professional fees	191
15	Construction indices	201
16	Property insurance	205

| Index | | 207 |

Foreword

Estimates of the cost are needed at every stage through which construction projects pass. It is a worthy aim in construction project management that all significant decisions about design, construction and commissioning should be accompanied by an estimate of the cost of each of the options being considered.

One paradox of construction project management is that the earlier the decision has to be made, the more difficult is the task of estimating the cost implications. When design has been done and specifications, sizes and quantities are known, the cost estimating can be detailed and relatively accurate. When little or no design has yet been done, estimating is very much more difficult.

Another paradox of construction project management is that the most important decisions are taken earliest – when the cost estimates are the most difficult to obtain.

These considerations lead to the conclusion that cost estimating data arranged particularly to suit the very early stages of a project – when the budget is being set – will be particularly useful yet particularly difficult to obtain. The difficulty of obtaining the data may be the reason why no book devoted entirely to budget estimating in construction has been prepared until this one.

Bryan Spain's team has ranged widely to assemble the estimating data which this book contains. They have also thought carefully about the best way to arrange and present the data for budget estimating and written some purposeful and well informed commentary to help readers make best use of what is here.

Many projects will start on a firmer footing when the decision to go ahead is based upon the sort of estimate which this book will help managers to prepare.

Martin Barnes
Martin Barnes Project Management
Coopers and Lybrand Deloitte

Preface

This book has been written for everyone in the construction industry who has an interest in budget costs. All projects start (and sometimes die!) as an idea and invariably the first question to be asked is 'How much will it cost?' At this stage a wide variety of methods are used to assess the approximate cost of construction.

The book contains a broad range of information to help the developer, quantity surveyor, engineer, architect or landscape architect produce costings to enable decisions to be taken on the feasibility and financial viability of the proposed scheme. The book has been divided into classifications of work, i.e. building, civil engineering, mechanical and electrical, reclamation, landscaping, environmental work and refurbishment and most projects will contain an element of more than one classification.

Various terms are used to describe the result of the cost assessment, e.g. 'order of costs', 'approximate estimate', 'budget estimate' and others but they all have the same aim, to produce the most accurate financial forecast based upon the best information available. As further details are available the cost forecast should be upgraded and refined to match the changes so that the client has the best information available at any one time.

The authors have received help from many sources in the preparation of this book including the manufacturers of a wide range of products. Other valuable help was received from

Water Research Centre
Liverpool Polytechnic
Royal Institution of Chartered Surveyors
Building Cost Information Service
Association of Consulting Engineers
Royal Institute of British Architects
Association of Consultant Architects
Institute of Landscape Architects
Professor Michael C. Fleming
Paul Young
Dorothy Spain
Carol Downham

Our thanks are also due to John McGee of Martin Barnes Project Management (Coopers and Lybrand Deloitte) who wrote the chapter on Life Cycle Costing, Brendan O'Herlihy of Chestertons for the chapter on the Development Process and Andy Williamson of Building Services Design who contributed the cost information on Mechanical and Electrical work.

We are particularly grateful to Denys Milton who carried out most of the detailed research necessary for the production of the thousands of individual pieces of cost information that appear in the book.

We would welcome constructive criticism of the book together with suggestions for improving its scope and contents.

Whilst every effort is made to ensure the accuracy of the information given in this publication, neither the authors nor the publishers in any way accept liability of any kind resulting from the use made by any person of such information. The prices in this book are based upon costs current in the first quarter of 1990.

Bryan J.D. Spain, FInstCES MACost E
Leonard B. Morley, DipQS FRICS MInstCES MACostE

SPAIN AND PARTNERS
Consulting Quantity Surveyors
Unit 9
SMM Business Park
Dock Road
Birkenhead
Merseyside
L41 1DT

Introduction

The purpose of this book is to help in the preparation of budget estimates at the early stages of a project. These budget costs must be established at the feasibility stage to assess the projects viability, and different methods can be used to achieve this.

For building work, using square metre prices is usually the first approach but the chapters on elemental costs and composite rates should provide the reader with enough information to enable him to upgrade the accuracy of the square metre prices where necessary.

Civil engineering, reclamation and landscaping work do not lend themselves to the square metre method so information on principal item and composite rates has been included. The scope and meaning of these and other terms is set out below:

Square metre prices These give a selection of average costs per square metre of buildings based on the total floor area measured inside external walls and over all stairwells, liftwells, internal walls and other voids. It should be mentioned that these rates should be used only as a guide showing differences between building costs rather than as a measure by which budget estimates could be calculated. Wherever possible approximate quantities should be taken off to calculate budget estimates. In building work the prices are inclusive of Preliminary and General Items but do not include external works, equipment, furniture or fees for professional services. Preliminary and General Items costs vary between 7% and 14% as a percentage size of the contract. Estimators sometimes include part of the Preliminaries cost in the individual rates and the remainder in the Preliminaries section of the Bill of Quantities according to the methodology used by the contractor. The actual value of the Preliminaries in a contract is seldom less than 10%.

Elemental analyses A selection of cost analyses of various types of buildings is stated which should help the reader in preparing budget estimates from sketch drawings.

Composite rates These rates are most useful in budget estimating when applied to approximate quantities for the various elements of the building.

Introduction

The rates also enable comparisons to be made between the cost of different materials. The wide variation in the choice of materials available should enable the budget estimate to meet both the client's cost requirements and the design criteria. In Civil Engineering Work Chapter 5, the rates pull together some of the items listed in Chapter 4, e.g. prices for concrete walls include formwork and reinforcement. This information will be useful when there is only time to prepare very approximate quantities. It will also be of help when comparing the costs of different sizes of walls.

Principal rates The most frequently measured items have been selected with approximate rates set against them. These items and values could be used in the preparation of budget estimates where there is enough time available to prepare and price rough quantities. In civil engineering work, the user must remember to include an allowance for Class A – General Items. This could be expressed as a percentage (say 25% to 40% depending upon the nature of the work) or could be assessed in detail by using the rates in Class A.

Project costs This information is the most approximate and gives an 'order of costs' for any particular project. It would normally be used when an overall idea of the costs was required rather than a budget estimate.

Examples in the use of the book are set out below using the information in the various chapters to calculate the budget cost of an entire project.

Example 1

Client requirements

1. Factory: single storey 2000 m^2
2. Office: separate, two storey 400 m^2
3. Site: 10,000 m^2
4. Location: East Anglia
5. Siteworks: sloping embankment on one side, 150 m of road from highway, car parking for 100 cars

From Chapter 1 – Building prices per square metre		£
Factory: single storey owners requirements	£/m^2	350–450
Office buildings: owner occupation	£/m^2	850–1000

Building cost

Factory 2000 m^2 × £350–£450	700–900,000
Offices 400 m^2 × £850–£1000	340–400,000
	£1,040,000.00
Total building cost range to	£1,300,000.00

From Chapter 10			£
Site clearance	10,000 m² @ 0.25		2500
Demolition brick building	5,000 m³ @ 0.70		3500
Earth moving to level site	1,000 m³ @ 4.10		4100
Hydraulic mulch seeding	2,000 m² @ £4000/ha		800
Planting	500 nr @ £10		5000
Shrubs	1,000 nr @ £3		3000
Fences chain link 1.6 m high	400 m @ £20		8000
Gates	1 set @ £450		450
Tarmac road 7.5 m	150 m @ £250		37500
Car park spaces	100 nr @ £750		75000
		139850
Drainage to main outlet	150 m of 300 mm Hepseal pipe @ £40		6000
3 Nr manholes @ £350		1050
Car parking drainage	300 m of 150 mm Hepseal pipe @ £20		6000
5 Nr manholes @ £150		750
Sundry drainage		500
Incoming services	Allowance for L.A. charges		1,700
		155,850
Design contingencies 10%		15,150
		£171,000

Total estimated cost of scheme	£
Factory and offices: Average of £1,040,000 and £1,300,000	1,170,000
Site works	171,000
Total Budget Estimate	£1,341,000
Allowances for Regional Factor × 1.03	1,381,230
Allowance for fees (Chapter 14) @ 13%	179,560
£1,560,790
Including contingencies, say	£1,600,000

Example 2

Client requirements

1. Office: city centre 5000 m² office accommodation plus car park
2. Site: 1100 m²
3. Location: London
4. Site work: existing premises demolished by vendor, adjacent property on one side to be underpinned by developer.

From Chapter 1 – Building prices per square metre

	£
Offices for owner occupation good standard	850–1000
Extra for 4–7 floors	200– 250

Assumed offices can be provided in six floors with basement car park and services on seventh floor.

	£
5000 m² @ £1000 offices	5,000,000
1000 m² @ £850 car park	850,000
850 m² @ £750 service floor	637,500
Extra for 4–7 floors 6850 m² @ £225	1,541,250
	£8,028,750
Allowance for underpinning 35 m @ £900	31,500
	8,060,250
Extra for difficulties due to confined site	150,000
Total budget estimate	£8,210,250
Allowance for Regional Factor × 1.22	1,806,255
	10,016,505
Allowance for fees (Chapter 14) @ 13%	1,302,146
	£11,318,651
Including contingencies, say	£11,500,000

The map overleaf shows the regional variations in building costs and is reproduced by kind permission of the Building Cost Advisory Service (BCIS).

Figure 1 Standard Statistical Region, showing Regional Factors based on national average = 1.00.

PART ONE

Building Work

1 Square metre prices
2 Elemental costs
3 Composite rates

1
Square metre prices

The following range of prices can be applied to the types of projects listed. They include an allowance for Preliminary and General items but exclude external works, fittings furniture and professional fees. LP and H stands for lighting, power and heating.

Cost per square metre

Factories

£

Simple single storey steel frame (excluding L P and H)	250-275
Single storey for letting (including L P and H)	300-350
Single storey owners requirements	350-450
Single storey heavy industrial use	450-550
Single storey complex special requirements	600-850
Two storey steel frame owner requirements	750-900
Two storey high technology with offices	850-1250
Two storey with air conditioned environment	1500-2000

Warehouses

Low rise (7m) drive in (excluding L P and H)		235-280
Medium rise (14m) loading bays, owners requirements		350-450
High rise (15-20m) loading bays, owners requirements		450-600
Air conditioned, cold rooms	Add	200-400

Office buildings

Offices for letting minimum standard	550-800
Extra for 4-7 floors	150-200
Extra for air conditioning	150-200
Offices for owner occupation good standard	850-1000
Extra for 4-7 floors	200-250
Extra for air conditioning	200-350
Offices for owner occupation high standard	1200-2000
Extra for over 10 floors	300-500

BUILDING WORK

Office buildings (cont'd)	Cost per square metre £
Office attached to factory buildings	500-650
Refurbishment of existing office buildings	475-700
Extra for air conditioning	150-300
Extra for prestige building	400-500

Commercial premises

Supermarket buildings shell only	400-600
Extra for fitting out (excluding shop fitters work)	450-550
Retail warehouse outlets for letting (excluding L P and H)	250-400
Shopping parades (flats over) for letting	450-600

Public service buildings

Fire, police and ambulance stations	500-1000
Hospitals (refer to cost limits of Health Authorities)	1000-1500
Doctors' clinics	550-750
Health centres	600-800

Leisure premises

Sports facilities	550-900
Swimming pools, standard	1000-1350
Extra for luxury facilities	300-400
Cinemas shell only	400-500
Community centres	550-800
Restaurants and public houses	800-1000

Educational premises

Schools (refer to cost limits set by Department of Education and Science)	550-850
Colleges (polytechnic and training)	600-900
University colleges	700-1000
Specialist buildings	1000-1750

SQUARE METRE PRICES

Houses and flats

	Cost per square metre £
Local Authority and Housing Association	
Bungalows	525-625
Houses – two storey	500-600
Deduct for semi-detached	50
Deduct for terraced	100
Deduct for three storey	25
Flats	475-625
Extra for lifts	50-100
Sheltered housing	525-650
Private developments	
Houses and bungalows	600-800
Deduct for semi-detached	200
Flats	500-1000

Hotels

Motels	600-800
Town – medium quality	750-1000
City centre – luxury quality	1500-2000

Transport facilities

Airport hangars	1000-1250
Airport passenger terminal buildings (local)	1200-1500
Bus stations	500-700

Car parking facilities

Multi level	240-350
Underground	350-600

Garages and showrooms

Car showrooms	400-800
Petrol service stations (excluding area of forecourts)	700-1000

2
Elemental costs

The following analyses are intended to give the reader a broad selection of the cost of different types of buildings showing the varying costs of the elements. They should be used with care when constructing a cost plan because various methods of construction can affect the costs per square metre considerably.

However it is obviously more accurate to build up the elemental costs than to apply a single all-in rate for the whole building. External works, where shown, will also affect the final cost per square metre more than many of the other elements and consideration of this item is particularly important. The following buildings are included:

Factory unit – for letting
              – owner occupied

Factory and office – owner occupied

Warehouse unit – for letting

Office – two storey

Institution office – six storey
                   – civic

Superstore – shell

Health centre

Sports/leisure centre and swimming pool

Public house

Day nursery

School – secondary
       – hall

Flat – housing association

Elderly persons residence – single storey
                          – two storey

Garage workshop and showroom

BUILDING WORK

FACTORY UNITS FOR LETTING

Gross floor area: 15,200m2

Element No.		Cost/m2 floor area £	% of cost	Total cost of element £
1	Substructure	46.25	14.72	702,953
2	Frame	37.27	11.86	566,572
3	Upper Floors	3.44	1.09	52,223
4	Roof	23.85	7.59	362,582
5	Staircases	4.39	1.40	66,661
6	External Walls	20.72	6.59	314,872
7	Windows and External Doors	11.52	3.67	175,104
8	Partitions and Internal Walls	10.95	3.48	166,440
9	Internal Doors	5.45	1.73	82,849
10	Wall Finishes	0.29	0.09	4,364
11	Floor Finishes	9.64	3.07	146,453
12	Ceiling Finishes	2.13	0.68	32,422
13	Fittings	0.06	0.02	912
14	Sanitary Fittings	4.78	1.52	72,630
15	Waste Soil and Overflow Pipes	1.86	0.59	28,210
16	Hot and Cold Water Services	1.79	0.57	27,145
17	Heating Installation	29.38	9.35	446,638
18	Ventilation Installation	11.62	3.70	176,577
19	Gas Services	-	-	-
20	Electrical Installation	18.05	5.74	274,362
21	Lift Installation	-	-	-
22	Special Services Installation	3.64	1.16	55,362
23	Builders work in connection with Services	2.62	0.84	39,890
24	Drainage	10.91	3.47	165,770
25	External Works	51.12	16.26	776,961
	TOTAL	£311.71	100%	£4,737,952

ELEMENTAL COSTS

FACTORY: OWNER OCCUPIED

Gross floor area: 1720m2

Element No.		Cost/m2 floor area £	% of cost	Total cost of element £
1	Substructure	61.67	16.31	106,076
2	Frame	33.94	8.98	58,374
3	Upper Floors	-	-	-
4	Roof	73.94	19.56	127,177
5	Staircases	-	-	-
6	External Walls	26.55	7.02	45,674
7	Windows and External Doors	17.97	4.75	30,873
8	Partitions and Internal Walls	8.83	2.34	15,187
9	Internal Doors	1.93	0.51	3,318
10	Wall Finishes	9.99	2.64	17,185
11	Floor Finishes	4.38	1.16	7,534
12	Ceiling Finishes	3.54	0.94	6,085
13	Fittings	1.14	0.30	1,954
14	Sanitary Fittings	4.88	1.29	8,393
15	Waste Soil and Overflow Pipes	1.64	0.43	2,827
16	Hot and Cold Water Services	3.69	0.98	6,355
17	Heating Installation	1.11	0.29	1,908
18	Ventilation Installation	3.87	1.02	6,657
19	Gas Services	-	-	-
20	Electrical Installation	13.67	3.62	23,512
21	Lift Installation	-	-	-
22	Special Services Installation	-	-	-
23	Builders work in connection with Services	-	-	-
24	Drainage	22.31	5.90	38,380
25	External Works	83.04	21.96	142,827
	TOTAL	£378.08	100%	£650,297

BUILDING WORK

FACTORY AND OFFICES: OWNER OCCUPIED

Gross floor area: 7090m2

Element No.		Cost/m2 floor area £	% of cost	Total cost of element £
1	Substructure	34.45	11.78	244,497
2	Frame	62.24	21.28	441,804
3	Upper Floors	3.43	1.17	24,354
4	Roof	24.70	8.44	175,298
5	Staircases	1.99	0.68	14,149
6	External Walls	14.29	4.88	101,395
7	Windows and External Doors	14.92	5.10	105,888
8	Partitions and Internal Walls	6.51	2.22	46,188
9	Internal Doors	3.38	1.16	23,981
10	Wall Finishes	3.48	1.19	24,718
11	Floor Finishes	6.11	2.09	43,374
12	Ceiling Finishes	2.08	0.71	14,789
13	Fittings	-	-	-
14	Sanitary Fittings	1.66	0.57	11,767
15	Waste Soil and Overflow Pipes	1.33	0.45	9,445
16	Hot and Cold Water Services	-	-	-
17	Heating Installation	22.75	7.78	161,501
18	Ventilation Installation	3.74	1.28	26,532
19	Gas Services	-	-	-
20	Electrical Installation	26.10	8.92	185,263
21	Lift Installation	-	-	-
22	Special Services Installation	-	-	-
23	Builders work in connection with Services	-	-	-
24	Drainage	11.62	3.97	82,460
25	External Works	47.74	16.32	338,826
	TOTAL	£292.51	100%	£2,076,229

ELEMENTAL COSTS

WAREHOUSE UNITS FOR LETTING

Gross floor area: 11,150m2

Element No.		Cost/m2 floor area £	% of cost	Total cost of element £
1	Substructure	49.07	19.47	547,102
2	Frame	62.97	24.99	702,135
3	Upper Floors (included in 2)	–	–	–
4	Roof	49.42	19 61	551,050
5	Staircases	1.46	0.58	16,245
6	External Walls)			
7	Windows and External Doors)	53.81	21.35	599,955
8	Partitions and Internal Walls)			
9	Internal Doors)	3.98	1.58	44,417
10	Wall Finishes	0.75	0.30	8,378
11	Floor Finishes	0.67	0.26	7,421
12	Ceiling Finishes	1.58	0.63	17,645
13	Fittings	0.12	0.05	1,284
14	Sanitary Fittings	–	–	–
15	Waste Soil and Overflow Pipes	0.50	0.20	5,525
16	Hot and Cold Water Services	–	–	–
17	Heating Installation	2.67	1.06	29,825
18	Ventilation Installation	–	–	–
19	Gas Services	1.95	0.77	21,730
20	Electrical Installation	–	–	–
21	Lift Installation	–	–	–
22	Special Services Installation	0.01	0.01	81
23	Builders work in connection with Services	–	–	–
24	Drainage	23.07	9.16	257,263
25	External Works (not included)	–	–	–
	TOTAL	£252.02	100%	£2,810,056

BUILDING WORK

OFFICES TWO STOREY

Gross floor area: 1500m2

Element No.		Cost/m2 floor area £	% of cost	Total cost of element £
1	Substructure	77.10	12.58	115,643
2	Frame	29.59	4.83	44,383
3	Upper Floors	36.90	6.02	55,344
4	Roof	40.47	6.61	60,710
5	Staircases	-	-	-
6	External Walls	58.95	9.62	88,423
7	Windows and External Doors	128.80	21.03	193,206
8	Partitions and Internal Walls	4.61	0.75	6,918
9	Internal Doors	10.40	1.70	15,605
10	Wall Finishes	17.31	2.83	25,962
11	Floor Finishes	23.69	3.87	35,528
12	Ceiling Finishes	14.66	2.39	21,991
13	Fittings	0.42	0.07	625
14	Sanitary Fittings	3.62	0.59	5,437
15	Waste Soil and Overflow Pipes	4.88	0.80	7,323
16	Hot and Cold Water Services	-	-	-
17	Heating Installation	35.40	5.78	53,094
18	Ventilation Installation	-	-	-
19	Gas Services	-	-	-
20	Electrical Installation	36.95	6.03	55,428
21	Lift Installation	-	-	-
22	Special Services Installation	1.10	0.18	1,643
23	Builders work in connection with Services	1.53	0.25	2,302
24	Drainage	7.58	1.24	11,366
25	External Works	78.62	12.83	117,925
	TOTAL	£612.57	100%	£918,857

ELEMENTAL COSTS

INSTITUTION OFFICES 6 FLOORS

Gross floor area: 4216m2

Element No.		Cost/m2 floor area £	% of cost	Total cost of element £
1	Substructure	86.72	10.45	365,598
2	Frame & Upper Floors	101.48	12.23	427,850
3	Roof	10.66	1.28	44,927
4	Staircases in frame	–	–	–
5	External Walls	86.61	10.44	365,133
6	Windows and External Doors	61.92	7.46	261,061
7	Partitions and Internal Walls	24.35	2.93	102,640
8	Internal Doors	23.43	2.82	98,760
9	Wall Finishes	42.21	5.09	177,968
10	Floor Finishes	43.01	5.18	181,315
11	Ceiling Finishes	41.57	5.01	175,275
12	Fittings	42.49	5.12	179,123
13	Sanitary Fittings	4.99	0.60	21,031
14	Waste Soil and Overflow Pipes	–	–	–
15	Hot and Cold Water Services	111.50	13.44	470,094
16	Heating Installation	–	–	–
17	Ventilation Installation	–	–	–
18	Gas Services	–	–	–
19	Electrical Installation	52.32	6.31	220,582
20	Lift Installation	31.31	3.77	131,989
21	Special Services Installation	6.43	0.78	27,120
22	Builders work in connection with Services	10.49	1.26	44,216
23	Drainage	9.44	1.14	39,820
24	External Works	38.60	4.65	162,732
	TOTAL	£829.51	100%	£3,497,233

BUILDING WORK

CIVIC OFFICE BLOCK AIR CONDITIONED

Gross floor area: 4250m2

Element No.		Cost/m2 floor area £	% of cost	Total cost of element £
1	Substructure	41.15	5.22	174,870
2	Frame	58.46	7.41	248,436
3	Upper Floors	35.82	4.54	152,218
4	Roof	32.20	4.08	136,848
5	Staircases	15.14	1.92	64,325
6	External Walls	51.86	6.57	220,392
7	Windows and External Doors	47.96	6.08	203,831
8	Partitions and Internal Walls	15.13	1.92	64,301
9	Internal Doors	23.33	2.96	99,145
10	Wall Finishes	11.62	1.47	49,381
11	Floor Finishes	25.79	3.27	109,612
12	Ceiling Finishes	14.72	1.87	62,564
13	Fittings	40.27	5.10	171.168
14	Sanitary Fittings	3.71	0.47	15,774
15	Waste Soil and Overflow Pipes	8.81	1.12	37,446
16	Hot and Cold Water Services	26.50	3.36	112,605
17	Heating Installation Air Conditioning System)			
18	Ventilation Installation)	182.06	23.08	773,755
	)			
19	Gas Services	0.36	0.05	1,515
20	Electrical Installation	64.25	8.14	273,077
21	Lift Installation	16.14	2.05	68,594
22	Special Services Installation	7.22	0.92	30,692
23	Builders work in connection with Services	8.08	1.02	34,343
24	Drainage	10.44	1.32	44,388
25	External Works	47.92	6.07	203,679
	TOTAL	£788.93	100%	£3,352,958

ELEMENTAL COSTS

SUPERSTORE SHELL

Gross floor area: 2625m2

Element No.		Cost/m2 floor area £	% of cost	Total cost of element £
1	Substructure	60.77	21.09	159,530
2	Frame	81.60	28.31	214,201
3	Upper Floors	3.20	1.11	8,393
4	Roof	78.99	27.41	207,346
5	Staircases	0.47	0.16	1,223
6	External Walls	-	-	-
7	Windows and External Doors	25.11	8.71	65,923
8	Partitions and Internal Walls	17.43	6.05	45,749
9	Internal Doors	-	-	-
10	Wall Finishes	1.24	0.43	3,255
11	Floor Finishes	1.48	0.51	3,895
12	Ceiling Finishes	2.74	0.95	7,204
13	Fittings	-	-	-
14	Sanitary Fittings)			
15	Waste Soil and Overflow Pipes)	1.26	0.44	3,308
16	Hot and Cold Water Services)			
17	Heating Installation	-	-	-
18	Ventilation Installation	-	-	-
19	Gas Services	-	-	-
20	Electrical Installation	-	-	-
21	Lift Installation	-	-	-
22	Special Services Installation	-	-	-
23	Builders work in connection with Services	-	-	-
24	Drainage	13.91	4.83	36,504
25	External Works (not included)	-	-	-
	TOTAL	£288.20	100%	£756,532

BUILDING WORK

HEALTH CENTRE

Gross floor area: 1400m2

Element No.		Cost/m2 floor area £	% of cost	Total cost of element £
1	Substructure	96.48	7.84	135,072
2	Frame	-	-	-
3	Upper Floors	13.97	1.14	19,554
4	Roof	68.90	5.60	96,467
5	Staircases	7.72	0.63	10,806
6	External Walls	58.16	4.73	81,423
7	Windows and External Doors	61.95	5.04	86,732
8	Partitions and Internal Walls	29.88	2.43	41,837
9	Internal Doors	51.49	4.19	72,083
10	Wall Finishes	34.12	2.77	47,767
11	Floor Finishes	34.47	2.80	48,262
12	Ceiling Finishes	15.31	1.25	21,440
13	Fittings	75.25	6.12	105,354
14	Sanitary Fittings	19.22	1.56	26,910
15	Waste Soil and Overflow Pipes	2.04	0.17	2,852
16	Hot and Cold Water Services	45.54	3.70	63,754
17	Heating Installation	117.90	9.59	165,066
18	Ventilation Installation	18.56	1.51	25,984
19	Gas Services	0.38	0.03	539
20	Electrical Installation	107.95	8.78	151,126
21	Lift Installation	15.85	1.29	22,186
22	Special Services Installation	71.27	5.79	99,771
23	Builders work in connection with Services	27.63	2.25	38,683
24	Drainage	59.49	4.84	83,287
25	External Works	196.32	15.96	274,845
	TOTAL	£1,229.86	100%	£1,721,800

ELEMENTAL COSTS

SPORTS/LEISURE CENTRE AND SWIMMING POOL

Gross floor area: 3360m2

Element No.		Cost/m2 floor area £	% of cost	Total cost of element £
1	Substructure	58.63	6.14	196,991
2	Frame	75.67	7.92	254,258
3	Upper Floors	9.51	1.00	31,960
4	Roof	64.71	6.77	217,412
5	Staircases	9.67	1.01	32,487
6	External Walls	28.25	2.96	94,904
7	Windows and External Doors	24.35	2.55	81,819
8	Partitions and Internal Walls	88.83	9.30	298,472
9	Internal Doors	18.75	1.96	63,014
10	Wall Finishes	28.41	2.97	95,449
11	Floor Finishes	50.59	5.29	169,979
12	Ceiling Finishes	35.90	3.76	120,630
13	Fittings	41.14	4.31	138,238
14	Sanitary Fittings	5.10	0.53	17,125
15	Waste Soil and Overflow Pipes	1.23	0.13	4,117
16	Hot and Cold Water Services)			
17	Heating Installation)	156.96	16.63	527,377
18	Ventilation Installation)			
19	Gas Services	–	–	–
20	Electrical Installation	69.96	7.32	235,080
21	Lift Installation	7.17	0.75	24,077
22	Special Services Installation	107.45	11.25	361,025
23	Builders work in connection with Services	8.14	0.85	27,367
24	Drainage	18.19	1.90	61,131
25	External Works	46.82	4.90	157,302
	TOTAL	£955.42	100%	£3,210,215

BUILDING WORK

PUBLIC HOUSE

Gross floor area: 630m2

Element No.		Cost/m2 floor area £	% of cost	Total cost of element £
1	Substructure	109.06	13.41	68,710
2	Frame	7.52	0.92	4,736
3	Upper Floors	6.42	0.79	4,047
4	Roof	73.51	9.04	46,314
5	Staircases (included in 3)	-	-	-
6	External Walls	39.90	4.91	25,137
7	Windows and External Doors	23.50	2.89	14,805
8	Partitions and Internal Walls	17.34	2.13	10,922
9	Internal Doors	22.05	2.71	13,891
10	Wall Finishes	31.38	3.86	19,772
11	Floor Finishes	34.41	4.23	21,676
12	Ceiling Finishes	25.08	3.08	15,800
13	Fittings	78.52	9.65	49,469
14	Sanitary Fittings	13.46	1.66	8,482
15	Waste Soil and Overflow Pipes)			
16	Hot and Cold Water Services)	66.93	8.23	42,164
17	Heating Installation)			
18	Ventilation Installation)			
19	Gas Services	-	-	-
20	Electrical Installation	53.27	6.55	33,561
21	Lift Installation	-	-	-
22	Special Services Installation	22.94	2.82	14,451
23	Builders work in connection with Services	-	-	-
24	Drainage	38.28	4.71	24,117
25	External Works	149.84	18.42	94,400
	TOTAL	£813.42	100%	£512,455

ELEMENTAL COSTS

DAY NURSERY - HIGH COST EXTERNAL WORKS

Gross Floor area: 440m2

Element No.		Cost/m2 floor area £	% of cost	Total cost of element £
1	Substructure	129.96	9.99	57,184
2	Frame	45.28	3.48	19,924
3	Upper Floors	-	-	-
4	Roof	123.70	9.52	54,430
5	Staircases	-	-	-
6	External Walls	30.08	2.31	13,237
7	Windows and External Doors	57.31	4.41	25,215
8	Partitions and Internal Walls	16.73	1.29	7,361
9	Internal Doors	45.99	3.53	20,234
10	Wall Finishes	34.34	2.64	15,110
11	Floor Finishes	45.39	3.49	19,974
12	Ceiling Finishes	22.83	1.76	10,047
13	Fittings	86.57	6.65	38,092
14	Sanitary Fittings	16.39	1.26	7,212
15	Waste Soil and Overflow Pipes	10.00	0.77	4,402
16	Hot and Cold Water Services)			
17	Heating Installation)	141.98	10.91	62,472
18	Ventilation Installation)			
19	Gas Services	-	-	-
20	Electrical Installation	96.96	7.45	42,661
21	Lift Installation	-	-	-
22	Special Services Installation	0.94	0.07	415
23	Builders work in connection with Services	9.43	0.72	4,149
24	Drainage	81.98	6.30	36,073
25	External Works	305.00	23.45	134,201
	TOTAL	£1,300.89	100%	£572,393

BUILDING WORK

SECONDARY SCHOOL

Gross floor area: 2000m2

Element No.		Cost/m2 floor area £	% of cost	Total cost of element £
1	Substructure	59.02	9.88	118,038
2	Frame	45.94	7.69	91,876
3	Upper Floors	17.48	2.93	34,958
4	Roof	21.70	3.63	43,394
5	Staircases	10.85	1.82	21,705
6	External Walls	46.56	7.79	93,120
7	Windows and External Doors	28.48	4.77	56,966
8	Partitions and Internal Walls	21.40	3.58	42,803
9	Internal Doors	17.95	3.01	35,907
10	Wall Finishes	10.45	1.75	20,903
11	Floor Finishes	21.47	3.59	42,949
12	Ceiling Finishes	10.52	1.76	21,043
13	Fittings	19.80	3.31	39,601
14	Sanitary Fittings	5.71	0.96	11,416
15	Waste Soil and Overflow Pipes	9.76	1.63	19,513
16	Hot and Cold Water Services	13.28	2.22	26,562
17	Heating Installation	30.56	5.12	61,118
18	Ventilation Installation	7.82	1.31	15,635
19	Gas Services	4.31	0.72	8,627
20	Electrical Installation	29.22	4.89	58,442
21	Lift Installation	20.90	3.50	41,802
22	Special Services Installation	0.53	0.09	1,060
23	Builders work in connection with Services	10.33	1.73	20,667
24	Drainage	15.33	2.57	30,657
25	External Works	118.00	19.75	236,010
	TOTAL	£597.37	100%	1,194,771

ELEMENTAL COSTS

SCHOOL HALL HIGH QUALITY

Gross floor area: 250m2

Element No.		Cost/m2 floor area £	% of cost	Total cost of element £
1	Substructure	76.51	10.38	19,128
2	Frame	–	–	–
3	Upper Floors	0.68	0.09	170
4	Roof	163.70	22.20	40,924
5	Staircases	–	–	–
6	External Walls	145.95	19.80	36,487
7	Windows and External Doors	72.26	9.80	18,064
8	Partitions and Internal Walls	31.51	4.27	7,879
9	Internal Doors	12.73	1.73	3,184
10	Wall Finishes	7.72	1.05	1,931
11	Floor Finishes	31.72	4.30	7,931
12	Ceiling Finishes	5.04	0.68	1,261
13	Fittings	4.74	0.64	1,186
14	Sanitary Fittings)			
15	Waste Soil and Overflow Pipes)	1.82	0.25	455
16	Hot and Cold Water Services	–	–	–
17	Heating Installation)			
18	Ventilation Installation)	62.55	8.48	15,637
19	Gas Services	–	–	–
20	Electrical Installation	47.83	6.49	11,958
21	Lift Installation	–	–	–
22	Special Services Installation	–	–	–
23	Builders work in connection with Services	10.39	1.41	2,597
24	Drainage	18.86	2.56	4,716
25	External Works	43.23	5.86	10,807
	TOTAL	£737.26	100%	£184,314

BUILDING WORK

FLATS FOR HOUSING ASSOCIATION

Gross floor area: 1465m2

Element No.		Cost/m2 floor area £	% of cost	Total cost of element £
1	Substructure	69.40	9.44	101,670
2	Frame	–	–	–
3	Upper Floors	40.25	5.48	58,964
4	Roof	46.35	6.31	67,910
5	Staircases	13.61	1.85	19,944
6	External Walls	60.18	8.19	88,158
7	Windows and External Doors	56.83	7.73	83,251
8	Partitions and Internal Walls	37.16	5.06	54,433
9	Internal Doors	41.27	5.62	60,461
10	Wall Finishes	37.30	5.07	54,639
11	Floor Finishes	24.53	3.34	35,939
12	Ceiling Finishes	17.17	2.34	25,148
13	Fittings	20.31	2.34	29,759
14	Sanitary Fittings	13.67	1.86	20,031
15	Waste Soil and Overflow Pipes	1.09	0.15	1,604
16	Hot and Cold Water Services	14.53	1.98	21,289
17	Heating Installation (included in Electrical)	–	–	–
18	Ventilation Installation	–	–	–
19	Gas Services	–	–	–
20	Electrical Installation	108.39	14.75	158,785
21	Lift Installation	27.95	3.80	40,944
22	Special Services Installation	–	–	–
23	Builders work in connection with Services	3.69	0.50	5,406
24	Drainage	34.12	4.64	49,984
25	External Works	67.12	9.13	98,327
	TOTAL	£734.91	100%	£1,076,646

ELEMENTAL COSTS

FLATS FOR HOUSING ASSOCIATION

Gross floor area: 1280m2

Element No.		Cost/m2 floor area £	% of cost	Total cost of element £
1	Substructure	43.57	7.51	55,770
2	Frame	–	–	–
3	Upper Floors	14.28	2.46	18,284
4	Roof	49.95	8.61	63,930
5	Staircases	6.37	1.10	8,152
6	External Walls	53.09	9.15	67,953
7	Windows and External Doors	46.16	7.96	59,089
8	Partitions and Internal Walls	48.55	8.37	62,139
9	Internal Doors	36.38	6.27	46,564
10	Wall Finishes	21.46	3.70	27,474
11	Floor Finishes	12.87	2.22	16,470
12	Ceiling Finishes	31.42	5.42	40,213
13	Fittings	12.62	2.18	16,157
14	Sanitary Fittings)			
15	Waste Soil and Overflow Pipes)	19.99	3.45	25,588
16	Hot and Cold Water Services)			
17	Heating Installation)			
18	Ventilation Installation)	39.61	6.83	50,704
19	Gas Services	–	–	–
20	Electrical Installation	24.56	4.23	31,435
21	Lift Installation	–	–	–
22	Special Services Installation	–	–	–
23	Builders work in connection with Services	–	–	–
24	Drainage	30.42	5.24	38,933
25	External Works	88.84	15.31	113,720
	TOTAL	£580.14	100%	£742,575

BUILDING WORK

ELDERLY PERSONS RESIDENCE SINGLE STOREY

Gross floor area: 1300m2

Element No.		Cost/m2 floor area £	% of cost	Total cost of element £
1	Substructure	62.22	7.42	80,889
2	Frame	-	-	-
3	Upper Floors	-	-	-
4	Roof	106.85	12.74	138,908
5	Staircases	-	-	-
6	External Walls	40.07	4.78	52,084
7	Windows and External Doors	50.03	5.96	65,040
8	Partitions and Internal Walls	19.24	2.29	25,013
9	Internal Doors	40.86	4.87	53,123
10	Wall Finishes	27.03	3.22	35,141
11	Floor Finishes	32.39	3.86	42,111
12	Ceiling Finishes	11.84	1.41	15,391
13	Fittings	64.49	7.69	83,835
14	Sanitary Fittings	8.30	0.99	10,795
15	Waste Soil and Overflow Pipes	0.78	0.09	1,020
16	Hot and Cold Water Services	34.88	4.16	45,344
17	Heating Installation	38.46	4.59	50,000
18	Ventilation Installation	8.45	1.01	10,981
19	Gas Services	1.44	0.17	1,867
20	Electrical Installation	58.55	0.98	76,119
21	Lift Installation	-	-	-
22	Special Services Installation	35.01	4.17	45,514
23	Builders work in connection with Services	34.61	4.13	44,995
24	Drainage	33.87	4.04	44,027
25	External Works	129.45	15.43	168,285
	TOTAL	£838.83	100%	£1,090,482

ELEMENTAL COSTS

RESIDENTIAL UNIT FOR ELDERLY PEOPLE – TWO STOREY

Gross floor area: 400m2

Element No.		Cost/m2 floor area £	% of cost	Total cost of element £
1	Substructure	52.64	6.79	21,058
2	Frame	–	–	–
3	Upper Floors	37.72	4.87	15,088
4	Roof	108.77	14.03	43,507
5	Staircases	14.27	1.84	5,708
6	External Walls	31.63	4.08	12,652
7	Windows and External Doors	69.20	8.93	27,682
8	Partitions and Internal Walls	11.17	1.44	4,469
9	Internal Doors	55.56	7.17	22,224
10	Wall Finishes	31.83	4.11	12,734
11	Floor Finishes	43.49	5.61	17,395
12	Ceiling Finishes	22.98	2.97	9,193
13	Fittings	49.97	6.45	19,989
14	Sanitary Fittings	10.18	1.31	4,071
15	Waste Soil and Overflow Pipes			
16	Hot and Cold Water Services	35.72	4.61	14,286
17	Heating Installation			
18	Ventilation Installation	49.13	6.34	19,654
19	Gas Services	0.87	0.11	347
20	Electrical Installation	25.63	3.31	10,252
21	Lift Installation	–	–	–
22	Special Services Installation	6.96	0.90	2,779
23	Builders work in connection with Services	6.06	0.78	2,424
24	Drainage	31.09	4.01	12,437
25	External Works	80.13	10.34	32,054
	TOTAL	£775.00	100%	£310,000

BUILDING WORK

GARAGE WORKSHOPS AND SHOWROOM

Gross floor area: 870m2

Element No.		Cost/m2 floor area £	% of cost	Total cost of element £
1	Substructure	72.84	10.14	63,370
2	Frame	112.81	15.70	98,148
3	Upper Floors	3.02	0.42	2,627
4	Roof (included in 2)	–	–	–
5	Staircases	–	–	–
6	External Walls	34.33	4.78	29,869
7	Windows and External Doors	68.36	9.52	59,469
8	Partitions and Internal Walls	40.04	5.57	34,837
9	Internal Doors	10.82	1.51	9,417
10	Wall Finishes	27.31	3.80	23,756
11	Floor Finishes	57.52	8.01	50,043
12	Ceiling Finishes	6.10	0.85	5,310
13	Fittings	4.78	0.67	4,157
14	Sanitary Fittings	–	–	–
15	Waste Soil and Overflow Pipes	–	–	–
16	Hot and Cold Water Services	51.28	7.14	44,617
17	Heating Installation	–	–	–
18	Ventilation Installation	–	–	–
19	Gas Services	–	–	–
20	Electrical Installation	51.99	7.24	45,228
21	Lift Installation	–	–	–
22	Special Services Installation	–	–	–
23	Builders work in connection with Services (included)	–	–	–
24	Drainage	53.81	7.49	46,814
25	External Works	123.28	17.16	107,254
	TOTAL	£718.30	100%	£624,918

3
Composite rates

When there is time to prepare approximate quantities to assess the budget value of a project, the following rates will be useful. They have been compiled by combining various items to produce composite rates.

SUBSTRUCTURES £

Excavate, dispose, uphold and prepare formation in basements	m3	15
Extra for		
made ground	m3	10-15
rock	m3	25-35
concrete or brickwork	m3	20-30
reinforced concrete	m3	30-35

Concrete (11.5N/mm2 40mm aggregate) in bases below formation including excavation, formwork and reinforcement, size

600 x 600 x 300 mm	nr	15-18
750 x 750 x 500 mm	nr	35-46
1000 x 1000 x 750 mm	nr	85-120
1200 x 1200 x 1000 mm	nr	200-230

Concrete (21N/mm2 20mm aggregate) in basement floor slab including blinding, Visqueen and reinforcement at 100 kg per m3, thickness

250 mm	m2	47-50
300 mm	m2	55-60
500 mm	m2	86-90

Concrete (21N/mm2 20mm aggregate) in basement walls including formwork and reinforcement at 115 kg per m3, thickness

250 mm	m2	93-100
300 mm	m2	100-115
500 mm	m2	133-145

BUILDING WORK

		£
Concrete (21N/mm2 20mm aggregate) in ground beams including excavation, formwork and reinforcement size		
500 x 500	m	65-70
750 x 500	m	87-90
900 x 600	m	127-130
1000 x 1000	m	205-210
Strip foundations for trench width 600mm including excavation disposal, earthwork support, concrete foundation (11.5 N/m2, 40 mm aggregate) 225 mm thick, common brick wall (£125 per thousand) height 600 mm, damp proof course		
215 mm thick	m	43-50
265 mm thick cavity wall	m	50-55
Strip foundations for trench width 600mm including excavation disposal, earthwork support, concrete foundation (11.5 N/m2, 40 mm aggregate) 225 mm thick, common brick wall (£125 per thousand) height 900 mm, damp proof course		
215 mm thick	m	65-70
265 mm thick cavity wall	m	75-78
Strip foundations for trench width 675mm including excavation disposal, earthwork support, concrete foundation (11.5 N/m2, 40 mm aggregate) 225 mm thick, common brick wall (£125 per thousand) height 600 mm, damp proof course		
327 mm thick	m	65-70
Strip foundations for trench width 675mm including excavation disposal, earthwork support, concrete foundation (11.5 N/m2, 40 mm aggregate) 225 mm thick, common brick wall (£125 per thousand) height 900 mm, damp proof course		
327 mm thick	m	95-100
Strip foundations for trench width 750mm including excavation disposal, earthwork support, concrete foundation (11.5 N/m2, 40 mm aggregate) 300 mm thick, common brick wall (£125 per thousand) height 600 mm, damp proof course		
450 mm thick	m	90-95

COMPOSITE RATES

		£
Strip foundations for trench width 750mm including excavation disposal, earthwork support, concrete foundation (11.5 N/m2, 40 mm aggregate) 300 mm thick, common brick wall (£125 per thousand) height 900 mm, damp proof course		
450 mm thick	m	127-133
Foundations for houses (measured over area of ground floor)		
Trenches 1.50 m deep filled with concrete, hardcore and concrete ground floor slab, thickened for internal walls	m2	100
Raft foundation for commercial development, low rise	m2	150
Underpinning adjoining property	m	900

<u>Piling</u>

See Civil Engineering Chapters 4 and 5.

BUILDING WORK

SUPERSTRUCTURES £

Frame and upper floors (Total of upper floor area)

Reinforced concrete frame and floors	m2	100-140
Extra for heavy floor loading	m2	20-30
Extra for large spans, normal loading	m2	15-35
Steel frame, concrete encased and floors	m2	125-175
Extra for heavy floor loading	m2	15-20
Extra for large spans, normal loading	m2	20-45

Industrial single storey (Ground floor area)

Precast concrete portal frame	m2	40-60
Structural steel portal frame (unencased)	m2	35-50
Structural steel portal frame (columns cased)	m2	50-80
Extra for large spans	m2	30-40

Floors only

Reinforced concrete suspended slabs	m2	40-60
Precast concrete suspended slabs	m2	25-40
Timber joists and chipboard	m2	20-30

Load bearing walls

Common bricks (£125 per thousand) in wall

102 mm thick	m2	25-28
215 mm thick	m2	45-49
327 mm thick	m2	67-69
440 mm thick	m2	80-85

Facing bricks (£300 per thousand) in wall

102 mm thick	m2	40-45
215 mm thick in two skins tied together	m2	80-85

Engineering bricks (£200 per thousand) in wall

102 mm thick	m2	30-32
215 mm thick	m2	57-62
327 mm thick	m2	82-90
440 mm thick	m2	110-117

COMPOSITE RATES

		£
Lightweight concrete blocks in wall		
75 mm thick	m2	13-15
90 mm thick	m2	15-18
100 mm thick	m2	16-19
140 mm thick	m2	20-24
190 mm thick	m2	24-29
Dense concrete blocks in wall		
75 mm thick	m2	16-18
90 mm thick	m2	19-21
100 mm thick	m2	20-23
140 mm thick	m2	25-28
190 mm thick	m2	30-34
215 mm thick	m2	33-37
Reconstructed stone blocks in wall		
100 mm thick	m2	35-40
Concrete (21N/mm2, 20mm aggregate) in wall including formwork and reinforcement		
100 mm thick	m2	50-54
150 mm thick	m2	55-60
200 mm thick	m2	65-68
250 mm thick	m2	70-79
300 mm thick	m2	75-86
Cavity wall formed with one leaf of common brickwork 102 mm thick (£125 per thousand) and one leaf of		
common brickwork (£125 per thousand) 102 mm thick	m2	56-60
facing brickwork (£300 per thousand) 102 mm thick	m2	73-76
engineering brickwork (£200 per thousand) 102 mm thick	m2	60-65
lightweight concrete block 100 mm thick	m2	48-52
dense concrete block 100 mm thick	m2	51-55
reconstructed stone block 100 mm thick	m2	68-72
in situ concrete wall 100 mm thick	m2	78-82

BUILDING WORK

Non-load bearing walls (internal finishes excluded)

		£
Precast concrete panels insulated, (with exposed aggregate finish add £50m2)	m2	100-110
Precast concrete with natural stone aggregate facing, (with relief features add £40-60m2)	m2	300-330
Profiled reinforced cement single skin sheeting on steel sheeting rails	m2	10-15
Extra for insulation and inner lining	m2	10-15
Profiled PVF2 coated galvanized steel sheeting	m2	25-35
Extra for insulation and inner lining	m2	10-15
Extra for full height block internal wall with decoration	m2	20-30
Fully insulated sandwich panels with PVF2 coated galvanized steel sheeting	m2	130-170
Curtain walling with galvanized steel members	m2	180-240
Curtain walling with anodized aluminium members	m2	200-250
Extra for double glazed units and sun filter glass	m2	50-80
Extra for high quality finish to aluminium	m2	200-400
Patent glazing in vertical cladding to industrial type buildings	m2	75-100
Extra for double glazed units	m2	65-80
Aluminium sheeting profiled and insulated with inner lining	m2	100-180
Extra over for finishes to walls		
two coat cement and sand with pea gravel rough cast	m2	8-10
two coat 'Tyrolean' finish	m2	10-12
one coat hardwall plaster 5 mm thick	m2	3-4
one coat hardwall plaster 13 mm thick	m2	4-6
two coat hardwall plaster 13 mm thick	m2	5-7
two coat lightweight plaster 10 mm thick	m2	4-6
tile hanging including battens and felt	m2	25-35
boarding in Western Red Cedar	m2	28-35
shingles in Western Red Cedar	m2	25-35
exposed aggregates finish in concrete	m2	5-15

COMPOSITE RATES

		£
Extra over finishes for walls (Cont'd)		
two coat emulsion paint	m2	3
one coat primer, one undercoat and one coat eggshell finish	m2	5
two base coats and coat 'Multicolour'	m2	6
lining paper (£0.75 per roll)	m2	2
woodchip paper (£1.20 per roll)	m2	3
flock paper (£6.50 per roll)	m2	4
Roof construction (no finishes, measured on plan)		
Reinforced concrete roof slabs	m2	37-50
Precast concrete roof slabs	m2	25-40
Softwood flat roofs	m2	28-40
Softwood trussed pitched roofs	m2	15-30
Steel trussed pitched roofs	m2	30-40
Flat roof decking (including finishes)		
Woodwool and three layer felt roofing	m2	35-50
Woodwool and asphalt two coat covering	m2	37-52
Extra for insulation upgrade	m2	4-10
Galvanised steel decking and three layer felt roofing	m2	40-53
Roof coverings comparative costs (Sloping, area measured on plan)		
Concrete interlocking tiles	m2	20-25
Clay pantiles	m2	25-40
Composition slates	m2	30-42
Natural slates (Welsh)	m2	55-65
Plain concrete tiles	m2	30-50
Plain clay tiles	m2	45-55
Handmade plain clay tiles	m2	70-90
Red cedar shingles	m2	60-65
Verges to pitched roof in softwood and plywood	m	15-20
Eaves to pitched roofs in softwood including PVC gutter	m	20-50

BUILDING WORK

		£
Ridge treatment		
concrete half round tiles	m	13-15
clay half round tiles	m	15-25
Hip treatment		
concrete half round tiles	m	15-17
clay half round tiles	m	20-25
bonnet hip tiles	m	35-45
Roof cladding comparative costs (sloping, area measured on plan)		
Non-asbestos profiled cladding	m2	14-16
Extra for coloured	m2	1-2
Extra for insulation panels	m2	12-15
Galvanized steel profiled cladding PVF2 coated	m2	20-30
Extra for insulation lining	m2	25-30
Extra for coloured lining	m2	2-5
Galvanized steel profiled cladding with insulated sandwich panels	m2	95-150
Aluminium profiled cladding PVF2 coated	m2	50-65
Extra for coloured insulation lining	m2	2-5
Aluminium profiled cladding with pre-painted finish	m2	25-40
Windows comparative costs (area of windows)		
Softwood standard windows painted, single glazed	m2	130-180
Extra for double glazed	m2	40
Softwood windows, purpose made, painted, single glazed	m2	180-240
Extra for double glazed	m2	50
Steel standard windows with galvanized finish painted and single glazed	m2	150-200
Extra for double glazed	m2	40
Extra for self-finish colour coated	m2	5
Steel windows, purpose made, colour coated	m2	200-250
Extra for double glazed	m2	50
Hardwood windows, purpose made, stained and single glazed	m2	250-350

COMPOSITE RATES

		£
Extra for double glazed	m2	50
UPVC windows purpose made	m2	400-500
Extra for sun-resistant glass	m2	25
Aluminium windows, purpose made, anodised, double glazed, hardwood frame	m2	300-500

External doors (including frames, ironmongery and finishes)

Softwood external standard panelled doors	nr	175-225
Extra for glazed panels in decorative glass	nr	100-150
Hardwood external panelled door	nr	400-500
Solid core flush external door - single	nr	250-300
Solid core flush external door - pair	nr	350-400
Extra for glazed panels	nr	20-30
Extra for panic bolts, double doors	nr	100-150
Steel faced security flush external door	nr	1000-1500
Steel roller shutters	m2	120-200
Extra for electric motor operation	m2	50-100
Entrance doors and screens	m2	500-1000
Revolving doors - purpose made	nr	20000-30000
Rubber double doors	nr	1000-2000
Flexible strip curtains	nr	200-400

Internal partition walls comparative costs

Half brick wall	m2	23-27
Lightweight blocks not exceeding 100mm thick	m2	13-15
Stud with plasterboard each side	m2	21-25
Proprietary manufacture plasterboard 65 mm thick	m2	25-30
Metal stud and plasterboard one hour resistance	m2	30-45
Extra for decoration	m2	2-6
Extra for flush doors and frames	nr	25-50
Demountable solid partitions with hardwood doors	m2	100-150
Permanent solid partitions with softwood doors	m2	50-100
Glazed aluminium partitions and doors	m2	75-110

BUILDING WORK

Demountable aluminium partitions and doors	m2	200-350
WC cubicles	nr	250-500

Internal doors (including frames, ironmongery and finishes)

Standard flush doors hollow core	nr	75-150
Extra for hardwood face	nr	10-20
Standard flush doors solid core	nr	200-250
Extra for double doors	nr	75-125
Extra for hardwood face	nr	10-20
Fire check doors standard finish	nr	125-200
Heavy duty flush doors, purpose made	nr	450-550
Extra for laminate finish	nr	100-120
Extra for overhead closer	nr	50-75
Standard panelled softwood doors	nr	200-350
Extra for hardwood	nr	100-125

Stairs

Reinforced concrete construction, 3250mm rise, granolithic finish

Straight flight, width

900mm	nr	1000
1200mm	nr	1200

Dog-leg flight, width

900mm	nr	1200
1200mm	nr	1500

Reinforced concrete construction, 3250mm rise, terrazzo finish

Straight flight, width

900mm	nr	2400
1200mm	nr	3000

Dog-leg flight, width

900mm	nr	2750
1200mm	nr	3350

COMPOSITE RATES

Softwood construction, 2600mm rise

    straight flight 900mm wide, no balustrade   nr   500-700

    two flights with quarter landing, softwood balustrade   nr   700-800

    two flights with half landing, hardwood handrail   nr   800-900

Mild steel construction, 3000mm rise

    straight flight 900mm wide   nr   2000-2500

    two flights with quarter landing   nr   2250-2750

    two flights with half landing   nr   2500-3000

Add for additional 300mm rise   nr   200-250

Mild steel spiral construction, 2000mm diameter, 3000mm rise

    perforated treads, no risers, including balustrade   nr   1500-2500

Extra for heavy duty stair   nr   500-1000

Balustrades, 3250mm rise

Mild steel

    straight flight   nr   900

    dog-leg flight   nr   1200

Stainless steel

    straight flight   nr   1500

    dog-leg flight   nr   2000

<u>Wall finishes</u>		Hardwall two coats £	Lightweight two coats £
Plaster and emulsion paint	m2	10-20	7-12
Extra for three coats oil paint	m2	2-3	
Extra for vinyl paper	m2	5-15	

		Thickness	
Plasterboard lining for direct decoration		9.5mm £	12.5mm £
with emulsion paint	m2	7-10	8-11
with plastic coating: insulated	m2	9-12	9-13

BUILDING WORK

Sheet linings on battens plugged to walls   £

Supalux 9mm thick	m2	13-16
Decorative plywood 6mm thick	m2	12-15
Extra for flame retardant boards	m2	5
Fibreboard and hardboards	m2	3-10
Asbestos - free boards	m2	8-18
'Formica' faced chipboards	m2	30-50
Gyproc wallboards 9.5mm thick finished with 'Drywall'	m2	7-11
Gyproc wallboards 12.5mm thick finished with 'Drywall'	m2	8-12
Extra for two layers 12.5mm	m2	6-7
Softwood boarding 19mm thick	m2	15-20
Softwood boarding 25mm thick	m2	20-25
Hardwood panelled linings	m2	120-150

Tile wall finishes

Ceramic tiles on backing of cement and sand	m2	25-50
High quality frost proof tiles with polysulphide mastic pointing	m2	50-80

Floor finishes

Board flooring

Softwood

tongued and grooved 25 mm thick	m2	10-15

Chipboard

tongued and grooved 22 mm thick	m2	7-9

Hardwood

tongued and grooved 22 mm thick	m2	25-50

Screeds

cement and sand	m2	5-10
latex cement	m2	4-6
granolithic	m2	10-15
epoxy floor finish	m2	20-25
extra heavy duty	m2	35-55

COMPOSITE RATES

		£
Tile paving		
quarry	m2	19-35
ceramic	m2	30-35
terrazzo	m2	40-80
vinyl	m2	10-40
cork	m2	15-25
carpet	m2	25-30
Wood block flooring		
herringbone pattern, sanded, wax polish		
sapele	m2	30-40
oak	m2	40-45
Sheet materials		
linoleum	m2	12-15
vinyl	m2	15-25
Extra for anti-static	m2	4-5
carpets including underlay	m2	15-30

<u>Ceiling finishes</u>

Concrete soffite		
hardwall plaster		
in one coat and emulsion	m2	5
in two coats and emulsion	m2	8
lightweight plaster in two coats and emulsion	m2	8
Plasterboard soffite		
'Thistle' skim coat and emulsion	m2	12
two coats of plaster	m2	3
two coats of oil paint	m2	3
textured plaster finish ('Artex')	m2	3
sprayed acoustic plaster	m2	15

BUILDING WORK

Suspended ceiling systems £

Standard quality, exposed flanges to grid	m2	23-25
Extra for acoustic tiles	m2	5-8
Extra for high quality tiles	m2	10-15
Aluminium suspended ceiling system - similar to 'Luxalon' with stove enamelled panels	m2	30-40
with 'egg crate' panels	m2	50-70
with patent decorative panels	m2	60-90

Sanitary fittings complete with waste pipework

Domestic

		Average White £	Coloured £
Ground floor			
lavatory basin with pedestal	nr	180	200
low level WC suite	nr	140	180
bath steel enamelled	nr	300	350
shower with tray and curtains	nr	375	400
stainless steel sink - double drainer	nr	200	250
First floor			
Extra for PVC stack and connections	nr		150-200
Extra for cast iron stack and connection	nr		300-400

Commercial

Lavatory basin in range	nr	200	250
Low level WC suite	nr	150	200
Bowl type urinal, fireclay	nr	300	320
Drinking fountain	nr	250	300
Shower tray and curtains	nr	375	400
Fireclay sinks for cleaners	nr	250	-
Slab urinals, 4 person	nr	850	-
Cast iron stack per floor	nr	100-200	-

Water services

Copper hot and cold water services to domestic or commercial sanitary fittings including all building work: Guide price £100-300 per point

For full details of mechanical and electrical services costs see Chapters 7 and 8.

COMPOSITE RATES

DRAINAGE

Pipe laying with allowance for machine excavation, disposal of surplus and fittings

Vitrified clay pipes laid in trench 500mm deep on 150mm granular bed and haunching, nominal bore £

100mm 'Supersleve' with push-fit joints	m	10-12
100mm 'Standard' pipes with mortar joints	m	12-14
Extra for concrete bed and haunching	m	3-4
150mm 'Hepsleve' with push-fit sleeve joints	m	14-16
150mm Hepseal with push-fit spigot and socket joints	m	17-20
Extra for 150mm concrete bed and surround	m	12-14
225mm 'Hepseal' pipes as before	m	25-27
Extra for concrete bed and haunch	m	5-7
300mm 'Hepseal' pipes as before	m	38-40
Extra for 150mm granular bed and surround	m	6-8

Cast iron pipes laid in trench 500mm deep on 150mm granular bed and haunching

100mm 'Time saver' mechanical joints	m	26-28
100mm Caulked lead joints	m	37-40
Extra for 150mm concrete bed and surround	m	10-12
150mm 'Time saver' mechanical joints	m	51-55
150mm Caulked lead joints	m	58-62
Extra for 150mm granular bed and surround	m	4-6

Add to foregoing rates for deeper trenches

1000 mm deep; pipes up to 200 mm bore	m	6-8
2000 mm deep; pipes up to 225 mm bore	m	18-20
3000 mm deep; pipes up to 300 mm bore	m	31-34

Add to foregoing rates for excavating by hand

500 mm deep; pipes up to 200 mm bore	m	4-6
1000 mm deep; pipes up to 200 mm bore	m	8-6
2000 mm deep; pipes up to 225 mm bore	m	20-25

BUILDING WORK

Manholes

£

Brick manhole including machine excavation, concrete base, engineering brickwork (class 'B') concrete cover slab, 600 x 600 cast iron medium duty cover, channels and benchings

600 x 450 x 1000 mm deep	nr	330-350
Extra for heavy duty triangular cover	nr	50-70
Add or deduct for depth variations per 100 mm		18-20
900 x 600 x 1500 mm deep	nr	575-600
Add or deduct for depth variations per 100 mm		25-30
1250 x 750 x 2000 mm deep	nr	850-900
Add or deduct for depth variation per 100 mm		35-40
1500 x 900 x 2500 mm deep	nr	1300-1400
Add or deduct for depth variations per 100 mm		40-50

Precast concrete ring manhole including machine excavation, concrete base, unreinforced concrete rings, heavy duty concrete cover slab with brickwork shaft 600 x 600 four courses high and heavy duty road manhole cover

675 mm shaft x 1000 mm deep	nr	275-300
Add or deduct for depth variation per 100 mm		7-10
900 mm shaft x 2000 mm deep	nr	440-460
Add or deduct for depth variations per 100 mm		10-12
1200 mm shaft x 3000 mm deep	nr	900-1000
Add or deduct for depth variations per 100 mm		25-30

Precast concrete ring manholes including machine excavation, concrete base, reinforced concrete rings, heavy duty concrete cover slab with brickwork shaft 600 x 600 four courses high and heavy duty road manhole cover.

1350 mm shaft	2000 mm deep	nr	750-800
Add or deduct for variation in depth 500 mm			110-125
1500 mm shaft	2500 mm deep	nr	1000-1200
Add or deduct for variation in depth 500 mm			140-160
1800 mm shaft	3000 mm deep	nr	1400-1600
Add or deduct for variations in depth 500 mm			175-200

COMPOSITE RATES

Inspection chambers

£

Cast iron inspection chambers with bolted access covers, caulked lead joints

100 x 100 with one branch	nr	110-130
100 x 100 with two branches one side	nr	200-220
150 x 100 with one branch	nr	150-160
150 x 150 with two branches one side	nr	340-360
150 x 150 with two branches each side	nr	389-420
225 x 150 with one branch	nr	500-600
225 x 150 with three branches	nr	600-650

Gullies including excavation and concrete

Concrete road gulley, heavy duty grating	nr	140-150
Stoneware yard gulley and grating	nr	70-80
Cast iron yard gulley and grating	nr	80-100
Rainwater gulley and grating	nr	25-35

Surface water drainage channels including concrete bed and surround; galvanised mild steel gratings accessories

150 mm wide	m	50-60
Extra for cast iron heavy duty grating	m	12-18
Extra for sump unit	nr	50-60
Extra for gulley assembly	nr	250-300

Pavings

Excavation and disposal to formation, 150 mm hardcore bed, blinding, compaction for surface paving

  plain concrete paths in bays: formwork to edges, tamped

100 mm thick	m2	15-20
150 mm thick	m2	18-25

  precast concrete paving slabs in cement lime mortar; painted

50 mm thick	m2	13-17
63 mm thick	m2	15-20

BUILDING WORK

		£
Tarmacadam on sub base 65 mm thick	m2	20-25

Roads

Estate roads including excavation and disposal to formation, base course sub base and wearing surfaces in two layers 75 mm thick bitumen macadam with 127 x 254 mm precast concrete kerbs each side

4.5 m wide	m	125-130
7.5 m wide	m	200-250
Extra for concrete roads 225 mm thick	m2	5-10

For further information see Chapters 9 and 10.

PART TWO

Civil Engineering Work

4 Principal rates
5 Composite rates
6 Project costs

4

Principal rates

GENERAL ITEMS (CLASS A - CESMM2)

The pricing of this section is highly subjective and the information set out below should be used with the knowledge that there are as many different approaches to the pricing of General Items and bill rates as there are civil engineering estimators.

Analysis of the General Items in tenders for civil projects show a proportion of between 15% and 40% of the total bill even when the overall level of the tenders are comparable.

The General Items listed below are based upon the needs of a civil project whose contract value is approximately £5m with an 18 month (78 weeks) contract period. The list of items given is not intended to be exhaustive.

Contractual requirements	£	£
Performance bond		
1% for construction period x 1.5 years x £3m	45,000	
¾% for maintenance period x 1 year x £3m	22,500	67,500
Insurance of the Works		
1½% x £3m	45,000	
Inflation say 1½% x 7½% x £3m	3,375	
Extra work say 1½% x 10% x £3m	4,500	52,875
Insurance of construction plant		
included in hire charges		nil
Insurance against damage to persons and property		
included in head office overheads		nil
Carried forward		£120,375

CIVIL ENGINEERING WORK

Brought forward		£120,375
Specified requirements		
Offices for Engineer's staff		
erect	250	
maintain and operate (100 weeks x £50)	5,000	
remove	250	5,500
Laboratory for Engineer's staff		
erect	200	
maintain and operate (50 weeks x £30)	1,500	
remove	200	1,900
Cabins for Engineer's staff		
erect	100	
maintain and operate (100 weeks x £20)	2,000	
remove	100	2,200
Services for the Engineer		
1800cc car (100 weeks x £110)	11,000	
landrover (78 weeks x £130)	10,140	
telephone, installation	150	
telephone, maintain and operate (100 weeks x £30)	3,000	24,290
Equipment for the Engineer		
office equipment (100 weeks x £50) comprising	5,000	
1 word processor		
2 desks		
2 tables		
1 conference table		
8 chairs		
2 filing cabinets		
sundries		
laboratory equipment (50 weeks x £50)	2,500	
surveying equipment (78 weeks x £40)	3,120	10,620
Carried forward		£164,885

PRINCIPAL RATES

Brought forward		£164,885
Attendance upon the Engineer		
driver (50 weeks x £240)	12,000	
chainman (78 weeks x £200)	15,600	
laboratory assistant, part time (30 x £300)	9,000	36,600
Testing of materials (included)		nil
Testing of works (included		nil
Temporary works		
traffic signals (30 weeks @ £15)	450	*
cleaning roads (40 weeks @ £300)	12,000	
progress photographs	1,000	
temporary lighting (40 weeks @ £40)	1,600	
temporary water supply, connection	1,000	
temporary water supply, pipework (200m x £3)	600	
temporary water supply (2,000,000 litres x 30p per 1000)	600	
temporary water supply, remove	200	
hardstanding (500m2 x £8)	4,000	
hardstanding, remove	500	21,950
Method Related Charges		
Offices for Contractor		
erect	350	
maintain and operate (100 weeks x £100)	10,000	
remove	300	
Cabins for Contractor		
erect	250	
maintain and operate (78 weeks x £40)	3,120	
remove	200	
Carried forward	£14,220	£223,435

CIVIL ENGINEERING WORK

		Brought forward	£14,220	£223,435
Stores for Contractor				
erect			250	
maintain and operate (100 weeks @ £40)			4,000	
remove			200	
Canteens and messroom for Contractor				
erect			400	
maintain and operate (78 weeks x £70)			5,460	
remove			250	
Electricity				
install			1,000	
maintain (100 weeks x £40)			4,000	
Plant (included)				
Supervision, administration				
agent		90 weeks @ £500 =	45,000	
assistant agents (2)		156 weeks @ £450 =	70,200	
inspectors (2)		156 weeks @ £350 =	54,600	
setting out engineer		30 weeks @ £400 =	12,000	
quantity surveyor		110 weeks @ £450 =	49,500	
section foremen (4)		312 weeks @ £350 =	109,200	
timekeeper/wages clerk		78 weeks @ £250 =	19,500	
storekeeper		78 weeks @ £250 =	19,500	
watchman		100 weeks @ £250 =	25,000	
tea boy		78 weeks @ £200 =	15,600	
Supervision, offloading and cleaning gang				
labourers (2) 140 weeks @ £250			35,000	484,880
		TOTAL		£708,315

PRINCIPAL RATES

GROUND INVESTIGATION (CLASS B - CESMM 2)

		Not in rock £	Partly in rock £	In rock £
Trial holes size 1 x 2m, depth				
not exceeding 1m	m	5	10	20
1 - 2 m	m	10	20	40
2 - 3 m	m	15	30	60
3 - 5 m	m	25	50	100
5 - 10m	m	50	100	200

Light percussion boreholes, 150mm diameter, depth		£
not exceeding 5 m	m	10
5 - 10 mm	m	20
10 - 20 mm	m	30
20 - 30 mm	m	45
30 - 40 mm	m	55

Rotary drilled boreholes, 150mm diameter without core recovery, depth		
not exceeding 5 m	m	10
5 - 10 mm	m	18
10 - 20 mm	m	25
20 - 30 mm	m	40
30 - 40 mm	m	50

Rotary drilled boreholes, 75mm diameter, with core recovery, depth	m	50
not exceeding 5 m	m	10
5 - 10 mm	m	55
10 - 20 mm	m	60
20 - 30 mm	m	65
30 - 40 mm	m	75

CIVIL ENGINEERING WORK

GEOTECHNICAL PROCESSES (CLASS C - CESMM2)

An allowance of £6,000 should be made for the establishment and removal of plant and equipment to carry out drilling operations.

	ne 5m £/m	5-10m £/m	10-20m £/m	20-30m £/m	30-40m £/m
Drilling for grout holes through material other than rock					
vertically downwards	12	15	17	20	22
downwards at an angle of 0-45° to the vertical	13	16	18	21	23
horizontally or downwards at an angle less than 45° to the horizontal	14	17	19	22	24
upwards at an angle 0-45° to the horizontal	18	22	24	27	30
upwards at an angle less than 45° to the vertical	20	24	26	30	32

	ne 5m £/m	5-10m £/m	10-20m £/m	20-30m £/m	30-40m £/m
Drilling for grout holes through rock					
vertically downwards	18	21	23	25	28
downwards at an angle of 0-45° to the vertical	19	22	24	26	29
horizontally or downwards at an angle less than 45° to the horizontal	20	23	25	27	30
upwards at an angle 0-45° to the horizontal	24	26	29	31	34
upwards at an angle less than 45° to the vertical	26	28	31	33	38

PRINCIPAL RATES

Grout materials and injections

An allowance of £6,000 should be made for the establishment and renewal of plant and equipment to carry out grouting operations.

£

Materials

cement	t	85
pulverized fuel ash	t	10
sand	t	13
pea gravel	t	12
bentonite	t	145

Injections

number of holes	nr	80

Diaphragm walls

It is assumed in the following rates that a minimum of 3000 m3 of excavation is required. An allowance of £40,000 should be made for the establishment and removal of plant and equipment to carry out the work.

Excavation in material other than rock or artificial hard material

maximum depth not exceeding 5m	m3	90
maximum depth 5-10m	m3	95
maximum depth 10-15m	m3	105
maximum depth 15-20m	m3	115

Concrete designed mix to BS5328; grade 20; ordinary portland cement to BS12; 20mm aggregate to BS882; walls 1000mm thick	m3	78

High yield bar reinforcement to BS4449

nominal size 12mm	t	590
nominal size 16mm	t	540
nominal size 20mm	t	510
nominal size 25mm	t	490
Waterproof joints	sum	200

CIVIL ENGINEERING WORK

		£
Concrete guide walls either side of excavation; each wall 1000mm wide x 500mm deep	m	130

Ground anchors

It is assumed in the following rates that a minimum of 75 nr anchors are to be installed and an allowance of £4,000 should be made for the establishment and removal of plant and equipment.

Ground anchorages, number in material other than rock to a maximum 10m depth; load 50 tonne

temporary	nr	90
temporary with single corrosion protection	nr	90
temporary with double corrosion protection	nr	90
permanent	nr	90
permanent with single corrosion protection	nr	90
permanent with double corrosion protection	nr	90

Total length of tendon in material other than rock

temporary	m	65
temporary with single corrosion protection	m	70
temporary with double corrosion protection	m	75
permanent	m	90
permanent with single corrosion protection	m	95
permanent with double corrosion protection	m	100

General anchorages, number in material which includes rock to a maximum 10m depth; load 50 tonne

temporary	nr	115
temporary with single corrosion protection	nr	115
temporary with double corrosion protection	nr	115
permanent	nr	115
permanent with single corrosion protection	nr	115
permanent with double corrosion protection	nr	115

PRINCIPAL RATES

		£
Total length of tendon in material which includes rock		
temporary	m	80
temporary with single corrosion protection	m	85
temporary with double corrosion protection	m	90
permanent	m	105
permanent with single corrosion protection	m	110
permanent with double corrosion protection	m	115

Sand, band and wick drains

It is assumed in the following rates that a minimum of 100 vertical drains are to be installed, and an allowance of £6,000 should be made for the establishment and removal of plant and equipment.

Number of drains	nr	60
Pre-drilled holes	nr	10
Depth of overlying material	m	6
Drains of maximum depth not exceeding 10m		
cross section 100-200mm	m	5
cross section 200-300mm	m	6
cross section 300-400mm	m	8
Drains of maximum depth 10-15m		
cross section 100-200mm	m	6
cross section 200-300mm	m	7
cross section 300-400mm	m	9
Drains of maximum depth 15-20m		
cross section 100-200mm	m	7
cross section 200-300mm	m	8
cross section 300-400mm	m	10

CIVIL ENGINEERING WORK

DEMOLITION AND SITE CLEARANCE (CLASS D - CESSM2) £

Site Clearance

General site clearance of areas

free from major obstructions	ha	600
woods, small trees and shrubs	ha	1500

Pull down trees (stumps measured separately) girth

500m-1m	nr	18
1-2m	nr	35
2-3m	nr	130
3-5m	nr	700

Grub up stumps and backfill with topsoil, diameter

150-500mm	nr	20
500mm-1m	nr	40
1-2m	nr	60

Demolish buildings

brickwork

volume 50-100m3	nr	200
volume 250-500m3	nr	1000
volume 1000-2500m3	nr	2000

concrete

volume 50-100m3	nr	300
volume 250-500m3	nr	1500
volume 1000-2500m3	nr	7000

Clay drains depth 1.5m including stone bed and surround	m	5
Concrete pipe depth 2m including concrete bed and surround	m	8

PRINCIPAL RATES

EARTHWORKS (CLASS E - CESMM2)

Dredging

It is extremely difficult to give even approximate cost information about excavation by dredging. The method used (cutter suction dredger, barge mounted excavator or grab hopper are some of the options), the depth of water and the disposal arrangements are all key factors.

The cost per cubic metre of dredging solid material should be in the range of £2 to £7 per cubic metre but specialist advice should be obtained even at the budget estimate stage.

		Average Condition £	Topsoil £	Stiff Clay £	Chalk £	Rock £
General excavation to reduce levels, maximum depth						
not exceeding 0.25m	m3	2.80	2.50	4.20	6.25	16.80
0.25-0.5m	m3	2.80	2.40	4.20	6.25	16.80
0.5-1m	m3	3.50	-	5.25	8.75	21.00
1-2m	m3	4.00	-	6.00	10.00	24.00
2-5m	m3	6.00	-	9.00	15.00	36.00
Excavation for foundation, maximum depth						
not exceeding 0.25m	m3	3.80	3.50	5.70	9.50	22.80
0.25-0.5m	m3	3.80	3.70	5.70	9.50	22.80
0.5-1m	m3	4.50	-	6.75	11.25	27.00
1-2m	m3	5.00	-	7.50	12.50	30.00
2-5m	m3	6.00	-	9.00	15.00	36.00

Disposal of excavated material	Unit	£ p
deposited on site 100m distance	m3	3.50
deposited on site 300m distance	m3	5.00
deposited off site 1km distance including tipping fees	m3	8.00
deposited off site 5km distance including tipping fees	m3	10.00

CIVIL ENGINEERING WORK

		£ p
Filling material compacted in layers		
surplus excavated material	m3	4.00
imported topsoil	m3	11.00
imported granular material (DTp Type 1)	m3	16.00
imported granular material (DTp Type 2)	m3	14.00

PRINCIPAL RATES

CONCRETE WORK (CLASSES F AND G CESMM2)

		£
Mass concrete (1:3:6 40mm aggregate) supplied and placed in		
blinding	m3	55
foundation	m3	58
walls	m3	60
slabs	m3	58
Reinforced concrete (1:2:4 20mm aggregate) supplied and placed in		
foundation	m3	64
walls	m3	66
slabs	m3	64
suspended slabs	m3	64
columns	m3	68
beams	m3	68
Formwork, rough finish		
horizontal	m2	16
vertical	m2	20
curved	m2	25
Formwork, fair finish		
horizontal	m2	18
vertical	m2	22
curved	m2	27
Reinforcement, mild steel bars BS4449 diameter		
8mm	t	840
12mm	t	750
16mm	t	710
20mm	t	680
25mm	t	670
32mm	t	660
40mm	t	650

CIVIL ENGINEERING WORK

		£
High yield steel bars BS4461 diameter		
8mm	t	850
12mm	t	760
16mm	t	720
20mm	t	690
25mm	t	680
32mm	t	670
40mm	t	660
Steel fabric BS4483 weight		
3-4 kg/m2, C385	m2	3
4-5 kg/m2, B385	m2	4
5-6 kg/m2, B503	m2	5
6-7 kg/m2, A393	m2	6

Joints

		£
Open surface plain	m2	2
Open surface plain with cork filler, thickness		
10mm	m2	10
20mm	m2	18
25mm	m2	20
Formed surface plain	m2	12
Formed surface plain with cork filler, thickness		
10mm	m2	20
20mm	m2	28
25mm	m2	30

PRINCIPAL RATES

PRECAST CONCRETE (CLASS H - CESMM 2)

Prestressed pre-tensioned members, concrete (1:2:4 20mm aggregate)   £

Beams size

100 x 65 x 1500mm	nr	10
200 x 65 x 2000mm	nr	20
250 x 100 x 2500mm	nr	30

Slabs, superimposed loads maximum 5 kg/m2, size

100 x 400 x 6000mm	nr	35
100 x 1200 x 6000mm	nr	80
100 x 2400 x 6000mm	nr	140
150 x 400 x 6000mm	nr	50
150 x 1200 x 6000mm	nr	110
150 x 2400 x 6000mm	nr	190

Copings, sills and weir blocks, weathered and throated, size

150 x 75mm	m	8
200 x 75mm	m	10
300 x 75mm	m	15

CIVIL ENGINEERING WORK

DRAINAGE (CLASSES I J K and L CESMM2)

Excavation

Excavating trenches for pipes, backfilling and removing surplus excavated material from site.

Excavation depth	Pipe diameter mm										
	150 £	225 £	300 £	375 £	400 £	450 £	525 £	600 £	900 £	1200 £	1500 £
500mm	4	5	-	-	-	-	-	-	-	-	-
750mm	7	8	9	-	-	-	-	-	-	-	-
1000mm	11	12	13	-	-	-	-	-	-	-	-
1250mm	13	14	14	-	-	-	-	-	-	-	-
1500mm	16	17	18	18	19	21	22	24	36	-	-
1750mm	19	20	21	21	24	25	26	28	38	-	-
2000mm	22	23	24	24	26	28	29	31	35	50	65
2500mm	30	31	32	34	35	37	38	39	43	55	75
3000mm	37	40	41	42	44	47	48	50	65	70	110
3500mm	45	48	50	51	52	57	58	60	72	82	125
4000mm	52	56	59	60	61	67	68	70	80	90	135
4500mm	68	70	73	74	77	84	85	86	95	105	115
5000mm	81	85	87	90	94	100	102	105	110	120	140
5500mm	100	102	104	107	110	115	117	120	130	140	170
6000mm	115	118	120	124	128	132	136	140	150	165	200

PRINCIPAL RATES

Beds, benching and covers

£ p

Sand bed to pipe nominal bore

150mm	m	0.80
225mm	m	0.85
300mm	m	1.00
375mm	m	1.05
400mm	m	1.10
450mm	m	1.20
525mm	m	1.30
600mm	m	1.45
900mm	m	1.90
1200mm	m	2.00
1500mm	m	2.90

Granular material, 10mm nominal size, bed to pipe nominal bore

150mm	m	1.90
225mm	m	1.00
300mm	m	1.10
375mm	m	1.15
400mm	m	1.30
450mm	m	1.35
525mm	m	1.45
600mm	m	1.60
900mm	m	2.15
1200mm	m	2.70
1500mm	m	3.25

Granular material, 10mm nominal size, bed and haunch to pipe nominal bore

150mm	m	1.35
225mm	m	1.70
300mm	m	2.10
375mm	m	2.90

CIVIL ENGINEERING WORK

Granular material 10mm nominal size, bed and haunch to pipe nominal bore (cont'd)

		£ p
400mm	m	3.15
450mm	m	3.40
525mm	m	3.80
600mm	m	4.25
900mm	m	6.70
1200mm	m	9.70
1500mm	m	12.90

Granular material, 10mm nominal size, bed and surround to pipe nominal bore

150mm	m	2.40
225mm	m	3.10
300mm	m	3.85
375mm	m	4.80
400mm	m	5.25
450mm	m	5.80
525mm	m	7.10
600mm	m	8.00
900mm	m	12.60
1200mm	m	18.50
1500mm	m	24.75

Concrete, mix 11.5 N/mm2 40mm aggregate, bed to pipe nominal bore

150mm	m	2.90
225mm	m	3.30
300mm	m	3.50
375mm	m	3.80
400mm	m	4.05
450mm	m	4.30
525mm	m	4.65
600mm	m	5.25
900mm	m	7.00

PRINCIPAL RATES

Concrete, mix 11.5 N/mm2 40mm aggregate, bed to pipe nominal bore (cont'd) £ p

1200mm	m	8.70
1500mm	m	10.45

Concrete, mix 11.5 N/mm2 40mm aggregate, bed and haunch to pipe nominal bore

150mm	m	4.30
225mm	m	5.50
300mm	m	6.85
375mm	m	9.30
400mm	m	10.15
450mm	m	11.00
525mm	m	12.20
600mm	m	13.75
900mm	m	21.50
1200mm	m	31.20
1500mm	m	41.50

Concrete, mix 11.5 N/mm2 40mm aggregate, bed and surround to pipe nominal bore

150mm	m	7.70
225mm	m	10.00
300mm	m	12.35
375mm	m	15.40
400mm	m	16.80
450mm	m	18.60
525mm	m	22.90
600mm	m	25.65
900mm	m	40.55
1200mm	m	59.50
1500mm	m	79.75

CIVIL ENGINEERING WORK

Pipes laid in trench £

Vitrified clay to BS65,
nominal bore

150mm	m	16
225mm	m	26
300mm	m	40
375mm	m	50
400mm	m	55
450mm	m	70

Ductile spun iron to BS4772 with Tyton joints,
nominal bore

150mm	m	20
225mm	m	40
300mm	m	55
375mm	m	75
400mm	m	80
450mm	m	90
525mm	m	115
600mm	m	120

uPVC to BS 3506

150mm	m	20
225mm	m	30
300mm	m	50

Concrete pipe to BS556, Class H,
nominal bore,

375mm	m	26
400mm	m	30
450mm	m	34
525mm	m	42
600mm	m	48
900mm	m	85
1200mm	m	130
1500mm	m	205

PRINCIPAL RATES

Concrete pipe to BS556,
Class M, nominal bore,
£

300mm	m	23
375mm	m	25
400mm	m	28
450mm	m	32
525mm	m	40
600mm	m	50
900mm	m	80
1200mm	m	125
1500mm	m	200

Concrete pipe to BS556,
Class L, nominal bore,

300mm	m	21
375mm	m	24
400mm	m	26
450mm	m	30
525mm	m	38
600mm	m	48
900mm	m	75
1200mm	m	115
1500mm	m	180

Manholes

	Depth						
	1m £	1.5m £	2m £	2.5m £	3m £	4m £	5m £
Brick manhole comprising in situ concrete base, channels and benching, precast concrete cover slabs, cast iron steps and medium duty cover size							
750 x 600mm	550	700	900	1100	–	–	–
1200 x 750mm	–	–	–	1300	1500	1800	2100

CIVIL ENGINEERING WORK

			Depth				
Precast concrete manhole comprising in situ concrete base, channels and benching precast concrete rings, cover slab, cast iron steps/ ladders and heavy duty cover, diameter	1m £	1.5m £	2m £	2.5m £	3m £	4m £	5m £
675mm	300	350	400	460	530	630	830
900mm	385	450	500	565	635	750	950
1050mm	450	600	650	750	825	950	1200
1200mm	-	750	700	850	1000	1250	1500
1500mm	-	900	1050	1200	1350	1475	1750
1800mm	-	1000	1175	1350	1600	2000	2500

Gullies

		£
Vitrified clay road gully, 480mm diameter x 900 mm deep, including excavation, concrete surround, Class B engineering brick seating and cast iron grating	nr	365
Precast concrete road gully, 375mm diameter x 900 mm deep, including excavation, concrete surround, Class B engineering brick seating and cast iron grating	nr	140

French drains

Fill French and rubble drains with Type A filter material	m3	17

Rectangular section ditches unlined cross sectional area

0.25-0.5m2	m	7
0.5-0.75m2	m	9
0.75-1m2	m	11
1-1.5m2	m	14
1.5-2m2	m	18

Rectangular section ditches lined with Filtram, cross sectional area

0.25-0.5m2	m	9
0.5-0.75m2	m	12
0.75-1m2	m	15

PRINCIPAL RATES

		£
Rectangular section ditches lined with Filtram, cross sectional area (cont'd)		
1-1.5m2	m	18
1.5-2m2	m	26
Trenches for pipes or cable not to be laid by the Contractor, cross sectional area		
0.25-0.5m2	m	6
0.5-0.75m2	m	8
0.75-1m2	m	10
1-5m2	m	13
1.5-2m2	m	16
One way cable ducts, 100m diameter in trench depth not exceeding		
1.5m	m	20
1.5-2m	m	27
2-2.5m	m	38
2.5-3m	m	42
Two way cable ducts, 100mm diameter in trench depth not exceeding		
1.5m	m	27
1.5-2m	m	34
2-2.5m	m	45
2.5-3m	m	50

Reinstatement

Breaking up and temporary reinstatement of trench for pipe nominal bore not exceeding 300mm		
grassland	m	4
road, 75mm thick bituminous macadam and base under	m	14
road, 100 mm thick bituminous macadam and base under	m	18

CIVIL ENGINEERING WORK

Breaking up and temporary reinstatement of
trench for pipe nominal bore not
exceeding 300mm (cont'd)   £

   road, 150mm thick bituminous macadam
   and base under   m   22

   pavement, 50mm thick bituminous macadam
   and base under   m   10

PRINCIPAL RATES

STRUCTURAL METALWORK (CLASS M – CESMM 2) £

Fabrication of members for frames

columns	t	840
beams	t	810
portal frames	t	800
trusses and built up girders	t	850
bracings, purlins and cladding rails	t	880
anchorages and holding down assemblies	t	920
permanent erection of members for frames	t	160

Site bolts

black	t	1600
HSFG general grade	t	1650
HSFG higher grade	t	2000
HSFG load indicating or load limit	t	2450

Offsite surface treatment

blast cleaning	t	70
wire brushing	t	102
galvanizing	t	165
one coat chromate primer	t	125

CIVIL ENGINEERING WORK

MISCELLANEOUS METALWORK (CLASS N CESMM 2) £

Stairways and landings	t	1850
Walkways and platforms	t	1600

Cat ladder in galvanized steel rungs at 300mm centres, strings extended to form handrail, 450mm wide, length

3000mm	nr	400
4000mm	nr	550
5000mm	nr	700
6000mm	nr	800
7000mm	nr	900
8000mm	nr	1050
9000mm	nr	1100
10000mm	nr	1300
Guard cage to cat ladder	m	35

Galvanized steel staircase 900mm wide with chequer plate treads balustrade one side, supported on universal columns

5500mm going, 3000mm rise, 16 treads and one landing	nr	2400
10000mm going, 5000mm rise, 16 treads and two landings	nr	3500
Galvanized tubular handrail, 1050mm high, standards at 2000mm centres, with middle rail	m	65
Galvanized flat section handrail and members, standards at 1000mm centres, infilled with square vertical bars at 100mm centres	m	100

Safety fencing

Tensioned corrugated to DTp Clause 409 with 'Z' section steel posts set in concrete

single side	m	30
double sided	m	45
Untensioned corrugated to DTp Clause 412, single sided with timber posts	m	30

PRINCIPAL RATES

Miscellaneous framing		£
Angle section		
150 x 75 x 10mm	m	12
100 x 100 10mm	m	11
150 x 150 x 12mm	m	15
Channel section		
150 x 75 x 10mm	m	13
250 x 75 x 16mm	m	17
Flooring		
Galvanized mild steel 'Durbar' pattern plate 8mm thick	m2	80
Galvanized open grid flooring 50mm thick	m2	145

CIVIL ENGINEERING WORK

TIMBER (CLASS O CESMM 2) £

Greenheart timber in marine works

100 x 75mm	m	16
150 x 75mm	m	22
200 x 200mm	m	65
200 x 300mm	m	70
300 x 300mm	m	95
600 x 600mm	m	275

Wrought softwood in marine work

100 x 75mm	m	8
150 x 75mm	m	10
200 x 200mm	m	18
200 x 300mm	m	30
300 x 300mm	m	38

Hardwood decking, thickness

50mm	m2	60
75mm	m2	75
100mm	m2	95

Softwood decking, thickness

50mm	m2	30
75mm	m2	35
100mm	m2	40

Coach screws, length

75mm	nr	2
100mm	nr	3
150mm	nr	4

Blackbolts, nuts and washers, length

100mm	nr	2
150mm	nr	3
200mm	nr	4

PRINCIPAL RATES

PILING (CLASS P - CESMM 2)

There are many different types of piling and the following have been included in this section

1. Bored cast in place concrete piles
2. Driven cast in place concrete piles
3. Pre-formed concrete piles
4. Isolated steel piles
5. Interlocking steel sheet piles

The costs are based upon constructing a minimum of 50 piles and include the establishing and dismantling of equipment separately.

Bored cast in place concrete piles

Cost of rig, set up and removal from site for £5500 for 50 piles, £10,000 for 100 piles.

		Diameter of pile mm			
		300	600	900	1200
Depth bored		£	£	£	£
10m	m	10	14	23	41
15m	m	9	13	21	37
20m	m	9	13	19	35
Concrete	m	13	28	52	96

Driven cast in place concrete piles

Cost of rig, set up and removal from site - £3250 for 50 piles, £6000 for 100 piles

Depth bored		300	600	900	1200
10m	m	8	10	13	18
15m	m	7	9	12	17
20m	m	7	9	12	17
Concrete	m	13	18	23	31

CIVIL ENGINEERING WORK

Preformed concrete piles

		Cross-sectional area mm		
		300x300	500x500	700x700
Design load	tonne	50	75	100
		£	£	£
Depth of pile				
10m	m	25	32	40
15m	m	22	30	38
20m	m	21	29	37

Add for transport of rig and piles £4000 - £5000 per job

Steel sheet piling

		Section modulus not exceeding cm3			
		700	1150	1700	2400
		£	£	£	£
Larsen/Frodingham sections	m2	85	100	115	130
extra for corners	m	28	28	31	33
Extra for junction	m	35	35	39	41

Add for transport of rig and piles, £3000-£4000 per job

Noise abatement requirements, £900-£1500 per job

PRINCIPAL RATES

ROADS AND PAVINGS (CLASS R - CESMM 2)   £

Sub-bases flexible road bases and surfacing

Hardcore road base depth 300mm	m3	13
Hardcore road base, depth		
100mm depth	m2	2
150mm depth	m2	3
Granular material, DTp type 1, depth		
100mm	m2	2
200mm	m2	4
300mm	m3	16
Granular material DTp type 2, depth 300mm	m3	14
Wet mix macadam DTp clause 808 base course, depth		
100mm	m2	3
200mm	m2	6
Dense bitumen macadam, DTp clause 908		
14mm aggregate, depth		
30mm wearing course	m2	2
40mm wearing course	m2	3
40mm aggregate, depth		
60mm base course	m2	4
80mm base course	m2	5

Concrete pavements

Carriageway slabs, grade C20, depth		
150mm	m3	60
150 - 350mm	m3	58
Carriageway slabs, DTp Specification C30 depth		
180mm	m2	14
250mm	m2	19

CIVIL ENGINEERING WORK

		£
Steel fabric reinforcement to BS4483		
reference A142 2.22 kgs/m2	m2	2
reference B503 5.93 kgs/m2	m2	3
Waterproof membranes below concrete pavements, plastic sheeting, 1200 gauge	m2	1

Joints in concrete pavements

Longitudinal joints, 10mm diameter x 750mm long mild steel dowels at 750mm centres, sealed with polysulphide, depth

150mm	m	7
220mm	m	9
250mm	m	10

Kerbs, channels and edgings   £

Precast concrete kerbs to BS340 straight or curved to radius exceeding 12m

fig 6 150 x 305mm	m	8
fig 7 125 x 255mm	m	7

PRINCIPAL RATES

RAIL TRACK (CLASS S - CESMM2)

Track foundations

		£
Bottom ballast granite, crushed, graded 50-25mm	m3	25
Top ballast granite, crushed, graded 50-25mm	m3	30

Taking up track and turnouts, dismantle and stack

Bullhead or flat bottom rail

plain track, timber sleepers, fishplate joints	m	4
turnouts, timber sleepers, fishplate joints	nr	125
extra for concrete sleepers	m	2

Sundries

buffer stops, weight 2-3 tonnes	nr	40

Lifting packing and slewing

Bullhead rail track on timber sleepers track length 10m maximum slew 300mm, maximum lift 300mm	nr	100
extra for concrete sleepers	nr	10
Turnout on timber sleepers	nr	300

Supplying only plain line material

Bullhead rails; for joints or welded track

mass 40 - 50 kg/m, section reference 95R	t	575

Sleepers

softwood timber, 250 x 125 x 2600mm long	nr	25
hardwood timber, 250 x 125 x 2600mm long	nr	35
concrete sleepers, type 'F27' with 2 nr cast iron 'Pandrol' fittings cast in	nr	35

Fittings

Chairs, cast iron 'CC' pattern complete	nr	25
Fish plates, standard set complete	nr	24
Fish plates, insulated set complete	nr	80

CIVIL ENGINEERING WORK

		£
Switches and crossings		
turnouts	nr	9000 - 14,000
diamond crossings	nr	8000 - 11,500
Sundries		
buffer stops, 2-2.5 tonnes	nr	1000 - 1,500

Laying only plain line material

Bullhead rails

plain track, mass 40-50 kg/m fish plate joints, timber sleepers	m	16
welded joints, concrete sleepers	m	30
form curve in plain track, raduis not exceeding 300mm	m	5
turnout, standard type on timber sleepers	nr	450
buffer stop, single rake	nr	125
'Thermit' weld on Bullhead rail	nr	85

PRINCIPAL RATES

TUNNELS (CLASS T - CESMM 2)

Excavation £

Tunnels in rock; straight

    diameter 1.8m | m3 | 100

Tunnels in clay; straight

    diameter 1.8m | m3 | 80

Shafts in rock; vertical

    diameter 3.0m | m3 | 80

Shafts in clay; vertical

    diameter 3.0m | m3 | 60

Other cavities in rock

    diameter 3.0m | m3 | 100

Other cavities in clay

    diameter 3.0m | m3 | 65

Excavated surfaces in rock

    cement grout | m2 | 10

Excavated surfaces in clay

    PFA/OPC grout (1:3) | m2 | 7

In situ lining to shafts; vertical

Cast concrete grade C20, primary

    diameter 2.13m | m3 | 85

Cast concrete grade C20; secondary

    diameter 2.13m | m3 | 90

Formwork; rough finish

    diameter 2.13m | m2 | 12

Formwork; smooth finish

    diameter 2.13m | m2 | 15

Pre-formed segmental linings to tunnels

Precast concrete bolted flanged rings

    diameter 2.74m special sealing gaskets | nr | 570

    diameter 3.05m standard gaskets | nr | 520

CIVIL ENGINEERING WORK

<u>Pre-formed segmental linings to shafts</u> £

Precast concrete bolted flanged rings

  diameter 4.27m   nr   800

  diameter 6.48m   nr   2000

PRINCIPAL RATES

BRICKWORK, BLOCKWORK AND MASONRY (CLASS U - CESMM2)   £

Common brickwork, BS3921, PC £125 per thousand in cement mortar (1:3)

  in vertical walls

102mm thick	m2	27
215mm thick	m2	43
327mm thick	m2	58
440mm thick	m2	78

  in columns and piers

215 x 215mm	m	12
327 x 327mm	m	26
440 x 440mm	m	45

Facing brickwork, PC £300 per thousand in coloured cement lime mortar (1:1:6)

  in vertical walls in Flemish bond

102mm thick	m2	40
215mm thick	m2	80

  in vertical curved walls in English bond

102mm thick	m2	50
215mm thick	m2	93
327mm thick	m2	139

Engineering brickwork, PC £200 per thousand in cement mortar (1:3)

  in vertical walls in English bond

215mm thick	m2	57
327mm thick	m2	96
440mm thick	m2	114

CIVIL ENGINEERING WORK

Engineering brickwork; PC £200 per thousand
in cement mortar

   in vertical facing to concrete

      102mm thick | m2 | 18

      215mm thick | m2 | 35

   in columns and piers

      215mm thick | m | 15

      327mm thick | m | 32

      440mm thick | m | 57

Blockwork; lightweight in cement mortar (1:3), stretcher bond

   in vertical straight walls

      100mm thick | m2 | 17

      140mm thick | m2 | 22

      190mm thick | m2 | 26

Blockwork; dense in cement mortar (1:3) straight walls

      100mm thick | m2 | 22

      140mm thick | m2 | 26

      190mm thick | m2 | 32

Ashlar masonry, Portland Whitbed with one exposed face in cement lime mortar (1:1:6) in vertical facing to concrete or brickwork

      100mm thick | m2 | 210

      200mm thick | m2 | 385

Rubble masonry, Cotswold limestone one exposed face, hammer dressed in cement lime mortar (1:1:6)

      200mm thick | m2 | 52

      350mm thick | m2 | 100

      500mm thick | m2 | 127

PRINCIPAL RATES

PAINTING (CLASS V - CESMM 2) £

(Rates inclusive of all inclinations)

One coat prime on general surfaces exceeding 300mm

metal	m2	2
timber	m2	2

Two coats of emulsion paint on general surfaces exceeding 300mm

smooth concrete	m2	2
blockwork and brickwork	m2	3

Two coats of cement paint

smooth concrete	m2	2
blockwork and brickwork	m2	3
rough cast surfaces	m2	4

Three coats of oil paint on

primed steel sections	m2	5
primed pipework	m2	6
planed timber	m2	5

CIVIL ENGINEERING WORK

WATERPROOFING (CLASS W - CESMM2) £

Damp proofing

One layer 1000 gauge 'Bituthene' sheet, fixed with adhesive	m2	6
Asphalt to BS1097 two coat work, 20mm thick on concrete surfaces	m2	12

Roofing

Asphalt to BS6577, two coat work, 20mm thick on concrete surfaces	m2	16
Built up felt roofing to BS747, three layer coverings	m2	12

Waterproofing

Proprietary roof decking with roof felt finish; insulated	m2	35
Protective layers, one layer 1000 gauge polythene sheet, fixed with adhesive	m2	1
Cement and sand (1:3) screed with waterproof-additive	m2	10
Sprayed or brushed waterproofing two coats of 'Synthaprufe' to concrete surfaces	m2	5

PRINCIPAL RATES

MISCELLANEOUS WORK (CLASS X - CESMM2)

£

Fences

Timber post and rail, driven posts height 1.10m with 4 rails	m	12
Timber round driven posts with 6 wires height 1.00m	m	6
Chain link fencing BS1772, with galvanized mesh, line and tying wire, concrete posts height 1.80m	m	19

Gates

Timber field gates BS3470, 1.10m high

softwood 2.70m wide	nr	75
hardwood 3.30m wide	nr	120

Gate posts 2.30m long

softwood 175 x 175mm	pair	125
concrete 200 x 200mm	pair	200

Drainage to structures above ground

UPVC half round system, to timber with brackets

gutters 110mm	m	6
gutters 120mm	m	7

Cast iron half round to timber with brackets £

gutters 125mm	m	12
gutters 150mm	m	16
Steel 'Plastisol' coated profiled to timber with over strap joints, 135 x 102mm	m	15

Rock filled gabions

Box type filled with graded broken rock

size 2000 x 1000 x 1000mm	nr	67
size 4000 x 1000 x 1000mm	nr	125

CIVIL ENGINEERING WORK

SEWER RENOVATION AND ANCILLARY WORKS (CLASS Y - CESMM2) £

Preparation of existing sewers

Cleaning egg shaped sewer 1050mm high	m	6-8
Removing intrusions		
brick	nr	3-5
laterals, clay bore not exceeding 150mm	nr	5-10
Plugging laterals		
bore not exceeding 300mm	nr	25-35
bore 750mm	nr	150-250
Local internal repairs		
areas 0.1-0.25 m2	nr	20-30
area 10 m2	nr	200-250

Stabilization of existing sewers

Pointing with cement mortar (1:3)	m2	15-20

Renovation of existing sewers

Segmental lining in GRP		
egg shaped 1050mm high	m	175-400
annulus grouting in Pozament	m3	120-150

Laterals to renovated sewers

Jointing		
bore not exceeding 150mm	nr	20-25
bore 150-300mm	nr	35-45
bore 450mm	nr	50-60

Interruptions

Preparation of existing sewers		
cleaning	hour	150-160
Stabilization		
pointing	hour	25-30
Renovation		
linings	hour	35-45

5

Composite rates

These composite items are mainly an amalgam of different item descriptions to assist in the quick preparation of approximate quantities. The rates represent the value of the combined scope of the items.

EARTHWORKS

Excavation £

General excavation of large area in good conditions

    extra for disposal on site up to 500m   m3   12

    extra for disposal off site up to 10 km   m3   15

Excavating foundations with JCB in good conditions

    extra for disposal on site up to 500m distance   m3   11

    extra for disposal off site up to 10 km distance   m3   14

Cutting for road or railway with scrapers in good conditions   m3   2

Filling and compacting excavated material in layers, large areas   m3   3

CONCRETE

Pile caps (21N/mm2, 20mm aggregate)

Rates inclusive of excavation, disposal, concrete, formwork, reinforcement at 110 kg per m3 and cutting away top of pile

Excavation 1500mm deep, concrete 1000mm thick

    1000 x 1000 mm   nr   324

    1500 x 1500 mm   nr   670

    2000 x 2000 mm   nr   1100

Excavation 2000mm deep, concrete 1200mm thick

    1000 x 1000 mm   nr   425

    1500 x 1500 mm   nr   875

    2000 x 2000 mm   nr   1450

CIVIL ENGINEERING WORK

Ground beams between pile caps (21N/mm2, 20mm aggregate)   £

Rates inclusive of excavation, disposal, concrete,
formwork and reinforcement at 140 kg per m3

500 x 500 mm deep	m	65
500 x 750 mm deep	m	96
600 x 1000 mm deep	m	145
750 x 1500 mm deep	m	171

CONCRETE STRUCTURES

Slabs on fill including blinding, Visqueen and
reinforcement at 100 kg per m3, thickness

150 mm	m2	32
200 mm	m2	40
225 mm	m2	45
250 mm	m2	48
300 mm	m2	57
500 mm	m2	88
Extra for waterproof concrete	%	10

Suspended slabs including formwork and
reinforcement at 150 kg per m3, thickness

150 mm	m2	44
200 mm	m2	53
225 mm	m2	57
250 mm	m2	61

Beams including formwork and reinforcement
at 240 kg per m3, size

250 x 300 mm	m	38
300 x 350 mm	m	48
350 x 450 mm	m	67
450 x 500 mm	m	87

COMPOSITE RATES

		£
Columns including formwork and reinforcement at 200 kg per m3, size		
250 x 250 mm	m	35
300 x 300 mm	m	45
450 x 500 mm	m	88
500 x 500 mm	m	95

Structural steelwork structures

Stanchions cased in concrete

254 x 254 x 132 kg/m	m	180
305 x 305 x 240 kg/m	m	325
356 x 406 x 467 kg/m	m	600

Stanchions cased in 'Supalux' board or similar material

254 x 254 x 132 kg/m	m	171
305 x 305 x 240 kg/m	m	309
356 x 406 x 467 kg/m	m	581

Beams cased in concrete

254 x 146 x 43 kg/m	m	69
305 x 165 x 54 kg/m	m	88
686 x 254 x 152 kg/m	m	235

Beams cased in 'Supalux' board or similar material

254 x 146 x 43 kg/m	m	65
305 x 165 x 54 kg/m	m	82
686 x 254 x 152 kg/m	m	220

CIVIL ENGINEERING WORK

Brickwork and blockwork

These rates are inclusive of forming openings, cutting, building in ties and pointing

		Thickness of wall		
		Half brick	One brick	Two brick
Common brickwork		£	£	£
vertical straight walls	m2	27	47	84
facing to concrete	m2	29	52	-
casing to steel columns	m2	35	56	-
Facing brickwork				
Bricks PC £250/1000				
vertical straight walls	m2	37	75	-
facing to concrete	m2	40	78	-
Bricks PC £350/1000				
vertical straight walls	m2	47	80	-
facing to concrete	m2	50	83	-
Engineering brickwork				
Staffordshire blue				
vertical straight walls	m2	45	80	150
piers and columns	m2	-	95	165
Accrington red				
vertical straight walls	m2	30	55	100
piers and columns	m2	-	70	125

		100mm Solid	140mm Solid	215mm Hollow
Dense concrete blockwork				
vertical straight walls	m2	17	22	28
casing to steel columns	m2	19	24	-

COMPOSITE RATES

Miscellaneous metal work   £

Steelwork staircases 3000 mm rise,

straight flight 1000 mm wide	each	nr	2250-2700
two flight with half landing	each	nr	3000-3500

Spiral staircase 2500 mm diameter, 3500 mm rise — nr — 2000-2250

Ladders in steel

cat ladder; 20 mm rungs at 300 mm centre, 470 mm wide, 3000 mm long	nr	350-400
extra for back hoops	nr	300-320
ships ladder; 75 mm wide treads at 300 mm centres, 470 mm wide, 3000 mm long	nr	450-500

Balustrades

balcony balustrades 835 mm high with 15 mm balusters at 200 mm centres 50 x 15 mm rounded handrails	m	65-75
staircase balustrade as last with ramps and wreaths	m	85-100
tubular steel balustrades 835 mm high with 40 mm diameter standards and handrail, galvanized	m	30-35

Open grid flooring

50 mm galvanized panels	m2	80-85
Durbar plate flooring 8 mm thick	m2	70-75

Plain steel sections, unfabricated beams, built into structure

457 x 191 x 82 kg/m UB	t	550
356 x 171 x 51 kg/m UB	t	570
254 x 146 x 37 kg/m UB	t	660

CIVIL ENGINEERING WORK

Drainage £

Pipe laying with allowance for machine excavation, disposal of surplus and fittings

Trenches 1000 mm deep, with granular bed and surround

ductile spun iron 150 mm	m	33
vitrified clay 150 mm	m	29
uPVC 150 mm	m	33

Trenches 1500 mm deep with concrete bed and haunch

ductile spun iron 225 mm	m	63
vitrified clay 225 mm	m	49
uPVC 225 mm	m	53

Trenches 2000 mm deep with concrete bed and surround

ductile spun iron 375 mm	m	114
vitrified clay 375 mm	m	89
concrete class H 375 mm	m	65

Trenches 2500 mm deep with granular bed and surround

ductile spun iron 400 mm	m	120
vitrified clay 400 mm	m	95
concrete class H 400 mm	m	70

Trenches 3000 mm deep with concrete bed and haunch

ductile spun iron 525 mm	m	175
concrete class H 525 mm	m	102

Trenches 4000 mm deep with granular bed and haunch

ductile spun iron 600 mm	m	195
concrete class H 600 mm	m	123

COMPOSITE RATES

AGRICULTURAL DRAINS DITCHES AND TRENCHES £

Porous stoneware or concrete in trenches
not exceeding 0.5m deep

75 mm	m	4
100 mm	m	6
150 mm	m	8

Trenches excavated and filled with rubble
not exceeding 0.5m deep

rectangular section 0.5m	m	14
vee section ditches unlined not exceeding 0.5m deep	m	9

Trenches for cable ducts (laid by others)
average 1000 mm deep

single way ducts not exceeding 200 mm bore	m	6
two way ducts	m	8

REINSTATEMENT OF SURFACING

80 mm tarmacadam surface with hardcore road base	m	13
150 mm concrete surface with hardcore road base	m	25

Manholes

Precast concrete ring manholes with reinforced concrete shaft rings, extra heavy duty concrete cover slab, brickwork shaft and 600 x 600 mm heavy duty road manhole cover

	Internal Diameter			
	1350mm £	1500mm £	1800mm £	2100mm £
2000 mm deep	775	1000	1150	1550
2500 mm deep	890	1150	1300	1800
3000 mm deep	1000	1300	1450	2000
3500 mm deep	1150	1450	1600	2250

Road gullies

Vitrified clay road gulley 450 mm diameter with heavy duty hinged cast iron grating, brickwork raising course	nr	250
Concrete road gulley 450 mm diameter and similar grating	nr	165

CIVIL ENGINEERING WORK

<u>Roadworks</u> £

Flexible surfaces

sub base in hardcore; thickness 150 mm	m2	3
road base in DTp Type 1; thickness 250 mm	m2	5

Dense bitumen macadam 40 mm aggregate

base course 60 mm	m2	5
wearing course 40 mm	m2	3
excavation and disposal 300-500 mm	m2	6
average for road construction as above	m2	26

Rigid surfaces

sub base in hardcore 150 mm	m2	3
road base in DTp Type 1 200 mm	m2	4
concrete surfacing with fabric reinforcement 150 mm	m2	12
excavation and disposal 300-500 mm	m2	6
average for road construction as above	m2	25

Kerbs and channels

precast concrete kerbs and foundations including excavation	m	11
extra for concrete channels	m	13

Tarmacadam road in two layers 80 mm thick, 250 mm road base on prepared formation including excavation, concrete kerbs each side

6.0 m wide	m	165
7.5 m wide	m	190

6
Project costs

The following information on budget estimating has been extracted from <u>Cost Information for Water Supply and Sewage Disposal</u> Technical report TR61 published by the Water Research Centre.

The report presents the results of a study, over many years, of costs of projects collected from local authorities and consultants. The results of the studies produced cost functions which can be updated by using indices.

The cost figures used are based on tenders received, not final costs, during the period 1960s to mid 1970s.

It must be emphasized that as with all statistically based, updated information, the results obtained from using the formulae should be treated with caution and the 'confidence limits' applied to the predictions. Reference to the original book will greatly assist the estimator to obtain more accurate results.

There are several sets of indices referred to in this chapter viz: The Construction Index, DQSD Index, Construction Materials Index, and Engineering and Allied Industries Index. If these indices are not readily to hand for updating purposes an indication of the increase from the quoted date may be obtained from the indices given in Chapter 15.

CIVIL ENGINEERING WORKS

SEWERAGE

$$\text{COST} = 0.000717 * \text{LEN}^{0.94} * \text{DIAM}^{0.72} * \text{DEP}^{0.57}$$

		Unit	Min.	Max.
where COST is total cost of scheme,		£'000 1976 Q3	2.5	2050
LEN is total length of pipework,		m	45	30 000
DIAM is mean diameter of scheme, weighting individual pipe diameters by their lengths,		mm	86	1440
and DEP is mean depth of scheme, weighting individual pipe depths by their corresponding excavated areas.		m	1.14	7.10

Number of cases: 80

Approximate multipliers for confidence limits about a prediction:

Confidence level	Lower	Upper
80%	0.56	1.78
95%	0.41	2.43

Note: 1. The New Construction Index should be used for inflation (value at 1976 Q3 = 246).

NB. See worked example on page 113

PROJECT COSTS

WATER PUMPING BUILDINGS

$$\text{WATPUMPCOS} = 4.00 * \text{THRUPUT}^{0.79}$$

	Unit	Min.	Max.
where WATPUMPCOS is total construction cost,	£'000 1976 Q3	49.2	874
and THRUPUT is design throughput.	'000 m3/d	15	680

Number of cases: 11

Approximate multipliers for confidence limits about a prediction:

Confidence level	Lower	Upper
80%	0.49	2.06
95%	0.31	3.26

Note: 1. The DQSD Index should be used for inflation (value at 1976 Q3 = 246).

NB. See worked example on page 113

CIVIL ENGINEERING WORKS

WATER MAINS

$$COST = 0.0702 * LEN^{0.73} * DIAM^{0.91}(DIAM/(1000 + DIAM))$$

		Unit	Min.	Max.
where COST	is total cost (i.e. installation and material cost) of scheme,	£'000 1976 Q3	70.3	4770
LEN	is total length of pipework,	m	744	45 500
and DIAM	is mean diameter of scheme, weighting individual pipe diameters by their lengths.	mm	126	1830

Number of cases: 37

Approximate multipliers for confidence limits about a prediction:

Confidence level	Lower	Upper
80%	0.64	1.55
95%	0.51	1.98

Note: 1. The New Construction Index should be used for inflation (value at 1976 Q3 = 246).

PROJECT COSTS

TUNNELS AND SHAFTS

$$\text{COST} = 0.0265 * \text{VOL}^{1.07}$$

	Unit	Min.	Max.
where COST is total tunnels and shafts cost of a scheme (see Note 2 below),	£million 1976 Q3	0.148	4.23
and VOL is the sum of the excavated volumes of the individual tunnels and shafts making up the contract.	'000 m3	4.75	131

Number of cases: 9

Approximate multipliers for confidence limits about a prediction:

Confidence level	Lower	Upper
80%	0.76	1.31
95%	0.64	1.57

Note: 1. The Construction Materials Index should be used for inflation (value at 1976 Q3 = 258).

2. COST is the cost of the individual tunnels and shafts making up the contract (assuming wedge-blocked lining), but excluding costs of secondary lining, shafts fittings, internal pipes and general and preliminary items.

3. Total contract cost may be estimated by first applying the above model and then multiplying by a LINING factor, which takes the value 1.43 for wedge-block lining, 1.57 for bolted concrete segment lining, and 2.00 for cast iron segment lining. This procedure is discussed more fully in Section 10.3D.

CIVIL ENGINEERING WORKS

INTAKES

Intake stations should be considered as pumping stations (with or without pumping plant) together with the additional bankside civil engineering and screening plant costs. Both of these additional items depend largely upon circumstances, but making simplifying assumptions the following table can be constructed.

£'000 1976 Q3

Throughput ('000 m3/d)	Pumping station		Intake station
	Building	Building and Pumping plant	Pumping station with intake structure and plant
2	6.92	13.5	19.6
5	14.3	27.6	41.4
10	24.7	47.4	69.8
20	42.6	81.4	120
50	88.0	167	230
100	152	286	382
200	263	491	640
500	542	1000	1310

Note:
1. The multipliers for confidence intervals about a prediction have been assumed to be similar to those given for the water pumping station building and pumping plant models; approximate values are as follows:

Confidence level	Lower	Upper
80%	0.5	2.0
95%	0.33	3.0

2. The New Construction Index should be used for inflation of civil engineering items (value at 1976 Q3 = 263).

3. The Engineering and Allied Industries Index should be used for inflation of plant items (value at 1976 Q3 = 227).

4. The figures assume a 50% standby pumping plant capacity.

5. Costs exclude any major interconnecting aqueduct between the intake and the pumping station.

PROJECT COSTS

SEWERAGE PUMPING PLANT

$$SEWCOS = 1.63 * CAP^{0.29} * HEAD^{0.19} * NPUMP^{0.89}$$

		Unit	Min.	Max.
where SEWCOS	is total plant cost,	£'000 1976 Q3	2.74	96.0
CAP	is total installed capacity (see Note 2 below),	l/s	1	1350
HEAD	is total head,	m	1.5	60.9
and NPUMP	is number of pumps installed.	–	1	7

Number of cases: 58

Approximate multipliers for confidence limits about a prediction:

Confidence level	Lower	Upper
80%	0.59	1.68
95%	0.45	2.23

Note: 1. The Engineering and Allied Industries Index should be used for inflation (value at 1976 Q3 = 227)

2. CAP is the combined operating and standby capacity (the extent of standby is decided by the user).

CIVIL ENGINEERING WORKS

SEWERAGE PUMPING BUILDINGS

$$COST = 6.97 * DESCAP^{0.21} * DESNPUMP^{0.60}$$

			Unit	Min.	Max.
where	PUMPCOS	is total construction cost,	£'000 1976 Q3	6.68	130
	DESCAP	is design capacity,	l/s	2	1440
and	DESNPUMP	is design number of pumps.	–	1	7

Number of cases: 58

Approximate multipliers for confidence limits about a prediction:

Confidence level	Lower	Upper
80%	0.58	1.74
95%	0.43	2.34

Note: 1. The New Construction Index should be used for inflation (value at 1976 Q3 = 263).

2. To estimate the cost of a complete pumping station, this digest should be used in conjunction with the sewerage pumping plant digest.

NB. See worked example on page 114

PROJECT COSTS

CONCRETE DAMS

$$\text{CONCOS} = 0.0569 * \text{DAMVOL}^{0.95}$$

	Unit	Min.	Max.
where CONCOS is total cost of dam (see Note 2 below),	£ million 1976 Q3	1.13	12.1
and DAMVOL is volume of fill of dam.	'000 m3	19	252

Number of cases: 13

Approximate multipliers for confidence limits about a prediction:

Confidence level	Lower	Upper
80%	0.77	1.29
95%	0.71	1.40

Note: 1. The Construction Materials Index should be used for inflation (value at 1976 Q3 = 258).

2. CONCOS includes the cost of the dam, cut-off, adjacent or integral inlet and outlet works, integral pipe and tunnel works, and minor diversions and ancillary works.

NB. See worked example on page 114

CIVIL ENGINEERING WORKS

EARTHBANK DAMS (with concrete cut-off walls)

$$\text{CONWALLCOS} = 8.97 * \text{DAMVOL}^{0.66}$$

		Unit	Min.	Max.
where CONWALLCOS	is total cost of dam (see Note 3 below),	£ million 1976 Q3	2.61	18.9
and DAMVOL	is volume of fill of dam, including all material placed and compacted.	million m3	0.116	3.00

Number of cases: 10

Approximate multipliers for confidence limits about a prediction:

Confidence level	Lower	Upper
80%	0.81	1.24
95%	0.70	1.42

Note: 1. The Construction Materials Index should be used for inflation (value at 1976 Q3 = 258).

2. The function applies only to earthbank dams constructed with a concrete cut-off wall that is substantial enough to act also as the core.

3. CONWALLCOS includes the cost of the dam, cut-off, adjacent or integral inlet and outlet works, integral pipe and tunnel works, and minor diversions and ancillary works.

NB. See worked example on page 114

PROJECT COSTS

EARTHBANK DAMS (with clay cores)

$$CLAYCORECOS = 4.53 * DAMVOL^{0.73} * TYPE^{-0.58}$$

			Unit	Min.	Max.
where	CLAYCORECOS	is total cost of dam (see Note 3 below),	£ million 1976 Q3	1.07	11.9
and	DAMVOL	is volume of fill of dam, including all material placed and compacted.	million m3	0.195	7.65
and	TYPE	is 2 for clay-cored bunds and 1 for other clay-cored dams.	Number of cases: 22		

Approximate multipliers for confidence limits about a prediction:

Confidence level	Lower	Upper
80%	0.82	1.22
95%	0.73	1.36

Note: 1. The Construction Materials Index should be used for inflation (value at 1976 Q3 = 258).

2. The effect on cost of using some rockfill or concrete grouting was statistically insignificant.

3. CLAYCORECOS includes the cost of the dam, cut-off, adjacent or integral inlet and outlet works, integral pipe and tunnel works, and minor diversions and ancillary works.

CIVIL ENGINEERING WORKS

RESERVOIRS AND LAGOONS

$$CLAYBUNCOS = 1.05*RESVOL^{0.68}$$

	Unit	Min.	Max.
where CLAYBUNCOS is total cost of embankment (see Note 3 below),	£ million 1976 Q3	0.0135	11.9
and RESVOL is the storage volume.	million m3	0.00226	37.7

Number of cases: 13

Approximate multipliers for confidence limits about a prediction:

Confidence level	Lower	Upper
80%	0.78	1.29
95%	0.67	1.50

Note: 1. The Construction Materials Index should be used for inflation (value at 1976 Q3 = 258).

2. The function applies only to clay-cored totally bunded reservoirs and simple excavated and/or bunded lagoons.

3. CLAYBUNCOS includes the cost of the dam, cut-off, adjacent or integral inlet and outlet works, integral pipe and tunnel works, and minor diversions and ancillary works.

PROJECT COSTS

WHOLE WATER TREATMENT WORKS

Total capital cost of water treatment works (£'000 1976 Q3)

Throughput ('000 m3/d)	Rock and moorland raw water Type (iii)		Moorland raw water Type (iv)	Lowland raw water Type (v)
	Pressure filtration	Gravity filtration	Sedimentation-filtration	
2	233	259	-	-
5	408	441	585	599
10	614	673	924	917
20	915	1 040	1 460	1 410
50	1 700	1 880	2 680	2 440
100	2 800	3 120	4 210	3 970
200	-	4 890	6 910	6 220
500	-	9 610	13 200	12 000

Confidence level	Approximate multipliers for confidence limits about a prediction			
80% (Upper	1.33	1.23	1.17	1.17
(Lower	0.75	0.81	0.85	0.85
95% (Upper	1.56	1.37	1.28	1.28
(Lower	0.64	0.72	0.78	0.78

Note: 1. Although no single index is appropriate for all the water treatment cost components, the New Construction Index was chosen in more than half the cases and so could reasonably be used here for inflation (value at 1976 Q3 = 263).

2. Cost includes civil engineering and building costs, mechanical and electrical engineering costs and sludge process costs, and includes all costs relating to conditions of contract. Costs associated with additional processes (see Section 12.6) and extra items (see Section 12.8.3) are excluded.

3. Costs have not been estimated separately for treatment of groundwater (raw water Type i) or upland rock water (raw water Type ii).

CIVIL ENGINEERING WORKS

SERVICE RESERVOIRS

$$\text{COST} = 0.0636 * \text{CAP}^{0.64}$$

	Unit	Min.	Max.
where COST is total cost of (rectangular concrete-covered) service reservoir,	£ million 1976 Q3	0.034	1.46
and CAP is tank capacity.	'000 m3	0.340	114

Number of cases: 47

Approximate multipliers for confidence limits about a prediction:

Confidence level	Lower	Upper
80%	0.69	1.44
95%	0.57	1.76

Note: 1. The New Construction Index should be used for inflation (value at 1976 Q3 = 263).

PROJECT COSTS

WATER TOWERS

$$COST = 162 * CAP^{0.77} * TYPE^{-0.56}$$

	Unit	Min.	Max.
where COST is total cost,	£'000 1976 Q3	11.5	514
and CAP is tank capacity,	'000 m3	0.060	3.41
and TYPE is 1 for concrete water towers, and 2 for steel water towers.	Number of cases:	21	

Approximate multipliers for confidence limits about a prediction:

Confidence level	Lower	Upper
80%	0.70	1.43
95%	0.57	1.77

Note: 1. The New Construction Index should be used for inflation (value at 1976 Q3 = 263).

2. The overall height of tower was not a significant variable; this is probably because of the limited variation of heights within the sample. The mean sample height was 25.1 m.

The South East Essex
College of Arts & Technology
Carnarvon Road Southend on Sea Essex SS2 6LS
Tel: Southend (0702) 220400   Fax: Southend (0702) 432320

CIVIL ENGINEERING WORKS

WHOLE SEWAGE TREATMENT WORKS

River and estuarine discharge

Capital costs (£'000 1976 Q3)

Effluent standard (SS/BOD/ammoniacal nitrogen)

Dry weather flow ('000 m3/d)	River 10/10/10		30/20		Estuarine 150/200	
	Civil	Mech.	Civil	Mech.	Civil	Mech.
2.5	522	229	452	184	–	–
5	947	377	753	303	371	169
10	1060	563	889	391	588	328
20	1710	944	1410	654	838	350
40	2760	1430	2250	937	1410	604
80	4820	2620	3950	1780	2370	931
160	7960	4210	6480	2770	3880	1650

Confidence level — Approximate multipliers for confidence limits about a prediction

80% (Upper	1.12	1.19	1.12	1.19	1.14	1.20
(Lower	0.89	0.84	0.89	0.84	0.88	0.83
95% (Upper	1.18	1.30	1.18	1.30	1.21	1.32
(Lower	0.85	0.77	0.85	0.77	0.83	0.76

Note: 1. The civil engineering and the mechanical engineering costs have both been presented so that cost predictions can be corrected for inflation. The New Construction Index should be used for inflation of the civil engineering costs (value at 1976 Q3 = 263); for the mechanical engineering costs the Engineering and Allied Industries Index should be used (value at 1976 Q3 = 227).

2. The costs include sludge process costs and all costs relating to conditions of contract. Costs associated with contractors' overheads, optional equipment such as administrative and laboratory buildings, and work specific to the site such as access roads, are excluded.

NB. See worked example on page 115

PROJECT COSTS

EXAMPLES

The following demonstrates the use of some of the formulae to calculate various budget estimates.

1. Sewerage scheme

    $COST = 0.000717 * LEN^{0.94} * DIAM^{0.72} * DEP^{0.57}$

    Assume the scheme has the following pipes

    1. 1500m 500mm pipes
    2. 750m   300mm pipes
    3. 500m   225mm pipes
    4. 500m   150mm pipes
    5. 1000m 100mm pipes

LEN 4250m 0.94 = 2574
DIAM 225mm 0.72 =   49
DEP 210m   0.57 =   15
From the formula cost = 0.000717 x 2574 x 49 x 1.5 = 135.65 x 1000

Total construction cost = £135,650 3Q 1976 Base date

   = £474,770 1Q 1990 Estimate date

80% confidence level £265,871 - £845,090

2. Water pumping building

    $COST = 4.00 * THRUPUT^{0.79}$

    Assume design throughput 45,000 m3/day
    From formula
    Total construction cost = £4 x $45^{0.79}$

    = 80.928

    say £81,000 at 3Q 1976

    £292,150 at 1Q 1990

CIVIL ENGINEERING WORKS

3. Sewage pumping building

   $PUMPCOST = 6.97 * DESCAP^{0.21} * DESNPUMP^{0.60}$

   PUMPCOST = total construction cost 000's

   DESCAP = design capacity

   DESNPUMP = design number of pumps

   Assume design capacity = 300 l/s

   number of pumps = 3

   From formula

   Total construction cost = $6.97 \times 300^{0.21} \times 3^{0.60}$

   $= 6.97 \times 3.31 \times 1.93$

   $= 40.42$

   $= £40,420$ at 3Q 1976

   $= £145,926$ at 1Q 1990

4. Concrete dam

   $COST = 0.569 * VOL^{0.95}$

   Assume the volume of the dam 150,000 m3

   From the formula : VOL in 000's

   Total construction cost = $0.0569 \times 150^{0.95}$

   $= 0.0569 \times 116.76$

   $= £6.64$ million in 3Q 1976

   $= £23.98$ in 1Q 1990

   Say £24 million

5. Earthbank dam (Clay-cored bund)

   $COST = 4.53 * VOL^{0.73} * TYPE^{-0.58}$

   Assume the volume of the dam 150,000m3

   From the formula : VOL in 000's

   Total construction cost = $4.53 \times 0.15^{0.73} \times 2^{-0.58}$

   $= 4.53 \times 0.25 \times 0.67$

   $= 0.76$

   $= £758,775$ at 3Q 1976

   $= £2,739,178$ at 1Q 1990

PROJECT COSTS

6. Whole sewage treatment works

   For the purpose of calculating a 'global cost' for a whole sewage treatment works some basic criteria are required to be known.

   1) Dry weather flow in '000 m3/d

   2) Effluent standard (SS/BOD/ammoniacal nitrogen)

   3) River or Estuarine discharge

   From the table:-

   Assuming : 10,000m3 per day
              30/20 Standard
              River discharge

   From the table : Civil Costs   £889,000
                    Mechanical Costs   £391,000
                                       £1,280,000   at 1976 3Q

                    Probable cost   £4,500,000   at 1990 1Q

              80% Confidence level   £5,000,000 - £3,900,000

PART THREE

Mechanical and Electrical Work

7 Square metre prices
8 Principal rates

7

Square metre prices

The square metre rates given for the mechanical and electrical work are for typical buildings. The rates assume that all public services are available on site. Professional fees are also excluded.

The rates are to be applied to the total floor area of all storeys of the building. The area measured is that between the external walls without deduction for internal walls or staircases.

Ranges of prices given for each building are typical. Specific requirements of the building(s) being considered must be taken into account. If varying types of buildings are used on a project as a whole, then the rates may need to be adjusted to allow for common preliminaries, site set-up, administration costs, plant etc.

The prices do not include incidental builders' work, profit and attendance or $2\frac{1}{2}$% cash discount for the main contractor.

Payments to statutory authorities and public undertakings for service provision or work carried out by them have been excluded.

The headings A, B and C given, classify the buildings into.

A - Speculative

B - Purpose Built - Class 1

C - Purpose Built - Class 2

Within the purpose built heading, Class 1 relates to those buildings that the Owner/Developer dictates specific requirements but where the engineering content is of an average complexity. For Class 2, the building is also purpose built but the engineering content is above average complexity. Where rates are not included it is considered that this class is not applicable.

The rates do not allow for machinery, computer terminals, kitchen and laundry equipment, sanitary ware, sprinkler systems and the like.

MECHANICAL AND ELECTRICAL WORK

MECHANICAL INSTALLATION

£ per square metre

Industrial	A	B	C
Factories	12-32	18-35	35-55
Warehousing	11-30	16-33	-
Assembly and Machine Workshops	-	35-50	50-70
Garage/Showrooms	-	15-30	-
Transport Garages	-	15-25	-
Agricultural			
Barns, Sheds, Glass Houses	-	5-10	-
Animal breeding units	-	15-20	-
Commercial			
Shops	40-60	45-70	-
Offices	50-90	125-180	180-270
Hotels	-	60-100	100-180
Supermarkets	-	110-180	-
Department Stores	-	120-200	-
Telecommunications	-	150-200	200-250
Computer Installations	-	150-200	200-250
Community			
Community Centres	-	80-110	-
Ambulance and Fire Stations	-	100-150	-
Bus Stations	-	30-50	50-110
Police Stations	-	120-170	-
Prisons	-	-	140-200
Churches	-	55-90	-
Concert Halls/Theatres	-	100-140	140-200
Museums/Libraries/Art Galleries	-	100-150	150-180
Magistrates/County Courts	-	180-250	-
Crown/High Courts	-	-	240-280

SQUARE METRE PRICES

Residential

£ per square metre

	A	B	C
Dormitory Hostels	–	45-50	–
Estate/Sheltered Housing	–	28-40	–
Houses for Individuals	18-22	18-32	32-40

Education

	A	B	C
Middle/Secondary Schools	–	60-100	–
University Buildings			
Administration	–	75-95	–
Laboratory/Research	–	75-180	–
Halls of Residence	–	50-70	–

Recreation

	A	B	C
Sports Centres	–	70-100	–
Squash Courts	–	40-60	–
Swimming Pools	–	80-100	–

Medical Services

	A	B	C
Clinics and Health Centres	–	80-120	–
Homes for the Elderly	–	50-70	–
Disabled Accommodation	–	50-80	–
G.P. Surgeries	–	35-40	–
Dental Surgeries	–	35-50	–
General Hospitals	–	–	160-300
Teaching Hospitals	–	–	180-300
Hospital Laboratories	–	75-180	180-350

MECHANICAL AND ELECTRICAL WORK

ELECTRICAL INSTALLATION

Industrial

£ per square metre

	A	B	C
Factories	15-30	22-36	36-60
Warehousing	12-32	22-38	-
Assembly and Machine Workshops	-	35-40	50-70
Garages/Showrooms	-	12-28	-
Transport Garages	-	14-25	-

Agricultural

	A	B	C
Barns, Sheds, Glass Houses	-	5-10	-
Animal breeding units	-	15-22	-

Commercial

	A	B	C
Shops	30-45	40-50	-
Offices	55-80	90-140	140-180
Hotels	-	40-70	70-120
Supermarkets	-	60-90	-
Departmental Stores	-	70-100	-
Telecommunications	-	100-160	160-180
Computer Installations	-	100-160	160-200

Community

	A	B	C
Community Centres	-	60-80	-
Ambulance and Fire Stations	-	85-130	-
Bus Stations	-	30-40	40-50
Police Stations	-	90-140	-
Prisons	-	-	80-120
Churches	-	15-20	-
Concert Halls/Theatres	-	100-120	120-140
Museums/Libraries/Art Galleries	-	60-90	90-100
Magistrates/County Courts	-	80-140	-
Crown/High Courts	-	-	120-180

SQUARE METRE PRICES

Residential

£ per square metre

	A	B	C
Dormitory Hostels	–	25–40	–
Estate/Sheltered Housing	–	25–35	–
Houses for individuals	15–30	15–30	30–35

Education

	A	B	C
Middle/Secondary Schools	–	45–60	–
University Buildings			
Administration	–	55–80	–
Laboratory/Research	–	55–85	–
Halls of Residence	–	25–40	–

Recreation

	A	B	C
Sports Centres	–	25–30	–
Squash Courts	–	20–25	–
Swimming Pools	–	25–30	–

Medical Serivces

	A	B	C
Clinics and Health Centres	–	25–40	–
Homes for the Elderly	–	20–30	–
Disabled Accommodation	–	20–40	–
General Practice Surgeries	–	15–30	–
Dental Surgeries	–	15–30	–
General Hospitals	–	–	95–140
Teaching Hospitals	–	–	100–140
Hospital Laboratories	–	55–85	85–100

MECHANICAL AND ELECTRICAL WORK

MECHANICAL INSTALLATION BASIC SERVICES

Industrial

£ per square metre

Factories, Assembly and Machine Workshops	A	B	C
Heating and Ventilation	16.50	22.00	37.00
Compressed Air Installation	10.00	18.00	22.00
Water Services	3.60	3.60	5.00
Fire Protection	1.50	1.50	1.50

(Process steam and ventilation are excluded)

Agricultural

	A	B	C
Heating and Ventilation	-	2.00	5.00
Water Services	-	1.00	3.00

Shops/Stores

	A	B	C
Heating and Ventilation	40.00	-	-
Air Conditioning	-	160.00	-
Water Services	3.60	3.60	-
Fire Protection	3.80	3.80	-

Offices

	A	B	C
Heating and Ventilation	50.00	-	-
Air Conditioning	-	160.00	240.00
Water Services	10.00	11.50	12.50
Fire Protection	3.80	3.80	3.80

Telecommunications/Computer Installation

	A	B	C
Heating and Ventilation	-	20.00	25.00
Air Conditioning	-	160.00	190.00
Water Services	-	11.50	12.50
Fire Protection	-	3.80	3.80

SQUARE METRE PRICES

Community	£ per square metre		
	A	B	C
Heating and Ventilation	–	50.00	50.00
Air Conditioning	–	160.00	190.00
Water Services	–	9.00	10.50
Fire Protection	–	3.80	3.80
Residential			
Heating and Ventilation	12.00	15.50	23.00
Water Services	9.50	12.50	15.00
Education			
Heating	–	30.00	58.00
Ventilation	–	20.00	62.00
Water Services	–	9.00	35.00
Fire Protection	–	2.30	2.30
Recreation			
Heating	–	15.00	20.00
Ventilation	–	5.00	7.00
Air Conditioning	–	22.00	30.00
Water Services	–	35.00	37.00
Fire Protection	–	3.80	3.80
Medical Services			
Heating	–	50.00	60.00
Ventilation	–	20.00	70.00
Air Conditioning	–	30.00	140.00
Water Services	–	12.00	40.00
Fire Protection	–	3.20	3.20

(Medical gases, vacuum and other dedicated services are excluded).

MECHANICAL AND ELECTRICAL WORK

ELECTRICAL INSTALLATIONS BASIC SERVICES

Industrial

£ per square metre

Factories, Assembly and Machine Workshops	A	B	C
Mains power and Switch Gear (excludes production supplies)	11.50	12.80	24.10
Telephones	0.70	0.90	0.90
General Lighting including fittings	14.50	17.30	19.00
Emergency Lighting	3.00	4.10	4.10
Fire Protection	2.05	3.50	3.50
Communications	1.05	1.05	1.20

Agricultural

Mains Power and Switch Gear	-	1.00	2.60
Lighting including fittings	-	2.00	3.50

COMMERCIAL

Shops/Stores

Mains Power and Switch Gear	16.00	35.00	-
General Lighting including fittings	17.00	38.00	-
Emergency Lighting	3.50	4.90	-
Telephones	0.90	1.25	-
External Lighting	-	0.80	-
Public Address	-	1.00	-
Security	-	1.35	-
Fire Protection	5.10	5.10	-
Lightning Protection	1.10	1.10	-

SQUARE METRE PRICES

Offices

£ per square metre

	A	B	C
Mains Power and Switch Gear	33.80	40.00	83.75
General Lighting including fittings	25.20	30.50	60.00
Emergency Lighting	3.70	5.10	7.50
Telephones	0.90	1.25	4.30
External Lighting	0.30	0.85	1.40
Public Address	–	1.10	1.30
Security	–	1.35	4.90
Fire Protection	5.10	5.10	5.10
Lightning Protection	1.10	1.10	1.10

Telecommunications/Computer Installation

	A	B	C
Mains Power and Switch Gear	–	62.00	97.00
General Lighting including fittings	–	58.30	62.00
Emergency Lighting	–	5.10	7.50
Telephones	–	1.25	4.30
Security	–	1.35	4.90
Fire Protection	–	5.10	5.10
Lightning Protection	–	1.10	1.10

Community

	A	B	C
Mains Power and Switch Gear	–	38.20	67.35
General Lighting including fittings	–	27.80	58.75
Emergency Lighting	–	3.50	7.50
Telephones	–	1.25	4.30
Security	–	1.30	4.50
Fire Protection	–	5.10	5.10
External Lighting	–	0.75	1.25
Lightning Protection	–	1.10	1.10

MECHANICAL AND ELECTRICAL WORK

Residential

£ per square metre

	A	B	C
Mains Power and Switch Gear	11.20	12.75	14.40
Lighting (including fittings)	16.50	20.00	25.25
Security	-	6.25	8.60
Telephones	-	1.40	1.60

Education

	A	B	C
Mains Power and Switch Gear	-	16.50	23.70
General Lighting including fittings	-	21.05	25.60
Emergency Lighting	-	5.10	7.50
Telephones	-	0.80	0.90
Security	-	1.80	2.30
Fire Protection	-	5.10	5.10
External Lighting	-	0.70	1.20

Recreation

	A	B	C
Mains Power and Switch Gear	-	8.20	-
General Lighting	-	10.50	-
Emergency Lighting	-	3.50	-
Telephones	-	0.80	-
Security	-	1.10	-
Fire Protection	-	3.50	-
External Lighting	-	0.70	-

Medical Services

	A	B	C
Mains Power and Switch Gear	-	11.40	60.30
General Lighting including fittings	-	18.15	45.15
Emergency Lighting	-	3.10	5.70
Telephones	-	0.90	3.70
Security	-	0.90	1.10
Fire Protection	-	5.10	5.10
Lightning Protection	-	1.10	1.10

8
Principal rates

The rates given for the mechanical and electrical services are for typical installations using average numbers of fittings and joints in relation to straight runs of pipework, conduit, trays, or trunking. For example, in the case of an office block with structural columns projecting in from the external wall and with a low temperature hot water system and dado height trunking, the linear metre rates do not allow for all the additional bends or sets around the columns. The figures in this particular case would need to be adjusted.

Where ranges of prices are given, the specific requirements of the installation or plant items must be taken into account. Adjustments will also need to be made for the size of project undertaken. For example with a very small project the overheads will be greater in proportion to the pipe or conduit length installed than in the case of a larger project.

The rates for plant and equipment include for normal works testing, delivery to site, off loading, storage, distribution up to 100 metres, and installation costs. The rates do not allow for any associated scaffolding or the use of tressels during the installation nor for any associated builder's work, profit and attendance or $2\frac{1}{2}$% cash discount for the Main Contractor.

MECHANICAL AND ELECTRICAL WORK

MECHANICAL WORK
DISTRIBUTION PIPEWORK OUTSIDE PLANT ROOMS
Mild Steel Pipework

Mild steel tube to BS1387 with joints as listed with normal allowance for fittings, brackets (clips directly to pipework) and wastage. Rates assume average runs. Thermal insulation is priced elsewhere. Valves not included in the rates.

Screwed jointing and fittings, nominal diameter		Heavy Weight £/m		Medium Weight £/m
15mm	m	33.18	m	31.64
20mm	m	17.22	m	16.24
25mm	m	18.76	m	17.50
32mm	m	19.60	m	18.06
40mm	m	20.58	m	19.04
50mm	m	42.00	m	40.04
65mm	m	50.26	m	48.02
80mm	m	55.72	m	52.08
100mm	m	57.54	m	56.31
125mm	m	64.82	m	62.10
150mm	m	111.02	m	108.64

Extra over for malleable iron fittings to BS143, nominal diameter		Elbow £/nr		Tee £/nr
15mm	nr	7.25	nr	10.33
20mm	nr	9.66	nr	13.91
25mm	nr	11.30	nr	16.03
32mm	nr	13.40	nr	19.08
40mm	nr	15.82	nr	22.21
50mm	nr	19.36	nr	27.34
65mm	nr	25.73	nr	36.24
80mm	nr	31.45	nr	42.60
100mm	nr	47.08	nr	62.26
125mm	nr	74.64	nr	108.48
150mm	nr	108.71	nr	149.28

Weld jointing and fittings, nominal diameter		Heavy Weight £/m		Medium Weight £/m
15mm	m	37.24	m	35.42
20mm	m	20.44	m	19.32
25mm	m	23.66	m	20.86
32mm	m	23.94	m	21.28
40mm	m	24.50	m	21.70
50mm	m	52.92	m	49.00
65mm	m	65.10	m	58.66
80mm	m	69.86	m	59.92
100mm	m	76.10	m	68.04
125mm	m	94.08	m	82.08
150mm	m	112.70	m	94.08

PRINCIPAL RATES

Extra over for heavy weight weldable fittings to BS1965 including welded joints, nominal diameter		Elbow Long Radius 90° £/nr		Tee Equal £/nr
25mm	nr	13.79	nr	28.03
32mm	nr	16.01	nr	31.27
40mm	nr	18.77	nr	35.99
50mm	nr	24.19	nr	46.06
65mm	nr	33.70	nr	60.46
80mm	nr	41.44	nr	69.56
100mm	nr	53.83	nr	89.66
125mm	nr	71.54	nr	145.98
150mm	nr	89.92	nr	171.66

N.B. for sizes 15mm and 20mm assumed pulled bends with branch welds in lieu of tees.

Copper Pipework

Light gauge copper tube to BS2871 Part 1 Table X with joints as listed, with normal allowance for fittings, brackets (clips to insulation spaces for cold water services, clips to pipe for hot water services) and wastage. Rates assume average runs. Thermal insulation priced elsewhere. Valves not included in the measure.

Soft solder capillary joints, nominal diameter		CWS £		HWS £
15mm	m	17.50	m	16.80
22mm	m	15.54	m	12.60
28mm	m	18.20	m	16.38
35mm	m	21.42	m	19.60
42mm	m	29.96	m	28.00
54mm	m	33.04	m	31.78

Extra over for capillary fittings to BS864, nominal diameter		Coupling £		Elbow £		Tee £
15mm	nr	3.19	nr	3.32	nr	5.01
22mm	nr	3.74	nr	4.06	nr	6.01
28mm	nr	4.58	nr	5.21	nr	7.71
35mm	nr	5.74	nr	7.32	nr	12.10
42mm	nr	6.90	nr	9.39	nr	14.87
54mm	nr	9.67	nr	15.44	nr	21.48

Manipulative compression joints, nominal diameter		CWS £		HWS £
15mm	m	17.08	m	15.54
22mm	m	14.98	m	12.32
28mm	m	18.20	m	16.38
35mm	m	21.98	m	20.30
42mm	m	31.64	m	30.52
54mm	m	35.14	m	33.74

MECHANICAL AND ELECTRICAL WORK

Extra over for compression fittings to BS864, nominal diameter

	Coupling £		Elbow £		Tee £	
15mm	nr	3.08	nr	3.25	nr	4.89
22mm	nr	4.05	nr	4.34	nr	6.09
28mm	nr	5.54	nr	6.31	nr	8.55
35mm	nr	8.40	nr	9.80	nr	13.79
42mm	nr	10.35	nr	12.95	nr	19.54
54mm	nr	14.18	nr	19.15	nr	28.59

Bronze welded jointing, nominal diameter

	CWS £		HWS £	
67mm	m	95.16	m	93.84

Extra over for bronze welded fittings, nominal diameter

	Coupling £		Elbow £		Tee £	
67mm	nr	34.73	nr	48.19	nr	69.82

THERMAL INSULATION

Pipework

Pre-formed, rigid section mineral fibre sections, thickness as indicated, with foil faced finish, fixed with adhesive and aluminium bands at 300mm centres. All joints and ends vapour sealed, including all pipe fittings, flanges and valves, mitred bends and tees. Nominal diameter

	Thickness of insulation			
	25mm £		40mm £	
15mm	m	7.00	m	7.70
20mm	m	7.38	m	8.12
25mm	m	7.81	m	8.59
32mm	m	8.19	m	9.01
40mm	m	8.44	m	9.28
50mm	m	12.06	m	13.27
65mm	m	14.75	m	16.23
80mm	m	18.50	m	20.35
100mm	m	20.44	m	22.48
125mm	m	23.63	m	25.99
150mm	m	27.94	m	30.73

PRINCIPAL RATES

Pre-formed, rigid section mineral fibre sections, thickness as indicated, with scrim canvas finish, fixed with aluminium bands at 500mm centres, with 0.6mm thick aluminium sheet secured with rivets or self tapping screws, including all pipe fittings, flanges and valves, mitred bends and tees. Nominal diameter	Thickness of insulation			
	25mm £		40mm £	
15mm	m	19.33	m	22.23
20mm	m	20.22	m	23.25
25mm	m	21.25	m	24.44
32mm	m	22.21	m	25.54
40mm	m	23.10	m	26.57
50mm	m	27.90	m	32.09
65mm	m	31.62	m	36.36
80mm	m	38.98	m	44.83
100mm	m	42.50	m	48.88
125mm	m	47.42	m	54.53
150mm	m	53.95	m	62.04

Ductwork and Flat Surfaces

Mineral fibre rigid slab reinforced aluminium foil faced, fixed to ductwork with pins and adhesive, joints and ends sealed with adhesive tape. Thickness of insulation	Surface Area Covered £	
25mm	m2	14.10
40mm	m2	15.11
50mm	m2	16.01
Extra for 0.9mm thick hammer clad aluminium sheet secured with rivets and self tapping screws	m2	17.21

Mineral fibre matteress, reinforced, foil faced, fixed to ductwork with pins and adhesive, and secured with 25mm galvanized wire mesh netting, joints and ends sealed with self-adhesive tape. Thickness of insulation		
25mm	m2	9.83
40mm	m2	10.49
50mm	m2	10.96

MECHANICAL AND ELECTRICAL WORK

BOILERS

Gas-fired, low temperature hot water heating boiler plant of shell and tube construction comprising of:
2 No boilers each rated at 60% heating load capacity, complete with boiler mountings and instruments
1 No Shunt pump
2 No Variable temperature circulating pumps
1 No 3-port valve arrangement for compensated circuit
2 No Constant temperature circulating pumps
Boiler control panel
Field located automatic controls
Feed and expansion tank
2 No Twin walled, stainless steel flues and all necessary distribution pipework, fittings, valves, thermal insulation, open vent and cold feed, including delivery, setting to work and commissioning.

Installed Boiler Capacity	Total Cost
300 kw	£38,000-54,000
1000 kw	£56,000-71,000

Extra for pressurization equipment in lieu of feed and expansion tank, cold feed and open vent.

300 kw	£4,800
1000 kw	£6,700

Gas-fired, low temperature hot water heating boiler plant of shell and tube construction, comprising of:
2 No Boilers, each rated at 60% heating load capacity complete with boiler mountings and instruments
2 No Copper storage calorifiers
1 No Shunt pump
2 No Variable temperature circulating pumps
1 No 3-port valve arrangement for compensated circuit
2 No Constant temperature circulating pumps
2 No Hot water service circulating pumps
Boiler control panel
Field located automatic controls
Feed and expansion tank
2 No Twin walled, stainless steel flues and all necessary distribution pipework, fittings, valves, thermal insulation, open vent and cold feed, including delivery, setting to work and commissioning.

Installed Boiler Capacity	Total Cost
300 kw	£40,000-59,000
1000 kw	£60,000-77,000

Extra for pressurization equipment on lieu of feed and expansion tank, cold feed and open vent for both LTHW and HWS systems.

300 kw	£9,000
1000 kw	£10,900

PRINCIPAL RATES

Oil-fired, using 35 sec oil, low temperature hot water heating boiler plant of shell and tube construction comprising of:
2 No Boilers each rated at 60% heating load capacity complete with boiler mountings and instruments
2 No Oil storage tanks, gravity supply
1 No Shunt pump
2 No Variable temperature circulating pumps
1 No 3-port valve arrangement for compensated circuit
2 No Constant temperature circulating pumps
Boiler control panel including back end protection
Field located automatic controls
Feed and expansion tank
2 No Twin walled, stainless steel flues and all necessary distribution pipework, fittings, valves, thermal insulation, open vent and cold feed, including delivery, setting to work and commissioning.

Installed Boiler Capacity	Total Cost
300 kw	£41,000-59,000
1000 kw	£61,000-78,000

Extra for pressurization equipment in lieu of feed and expansion tank, cold feed and open vent.

| 300 kw | £4,800 |
| 1000 kw | £6,700 |

Oil-fired, using 35 sec oil, low temperature hot water heating boiler plant of shell and tube construction, comprising of:
2 No Boilers each rated at 60% heating load capacity complete with boiler mountings and intruments
2 No Oil storage tanks, gravity supply
2 No Copper storage calorifiers
1 No Shunt pump
2 No Variable temperature circulating pumps
1 No 3-port valve arrangement for compensated circuit
2 No Constant temperature circulating pumps
2 No Hot water service circulating pumps
Boiler control panel
Field located automatic controls
Feed and expansion tank
2 No twin walled, stainless steel flues and all necessary distribution pipework, fittings, valves, thermal insulation, open vent and cold feed, including delivery, setting to work and commissioning

Installed Boiler Capacity	Total Cost
300 kw	£46,000-67,000
1000 kw	£65,000-84,000

Extra for pressurization equipment in lieu of feed and expansion tank, cold feed and open vent.

| 300 kw | £9,000 |
| 1000 kw | £10,900 |

MECHANICAL AND ELECTRICAL WORK

Packaged water boiler welded construction complete with boiler mountings and instruments, insulated casing, burner unit, delivery and commissioning.

Fuel	Rating	£
Gas	300 kw	5,270
Gas	1000 kw	12,740
Oil (35 sec)	300 kw	4,214
Oil (35 sec)	1000 kw	10,150

CHILLED WATER PLANT

Chilled water plant comprising of:
2 No Packaged water cooled water chillers
2 No Water cooling towers
2 No End suction, direct driven condenser water circulating pumps
2 No End suction, direct driven, chilled water circulating pumps
Interconnecting pipework, valves, fittings, thermal insulation, automatic controls, anti-vibration mountings, feed tank and chemical water treatment, commissioning and regulating valve sets.

Cost per ton of cooling capacity £950-2000

Extra for use of dry coolers in lieu of water cooling towers.

Cost per ton of cooling capacity £160

DUCTWORK

The rates given include for the sheet metal ductwork and associated fittings as listed. All labour and materials for the manufacture, delivery and erection for the fabrication of fittings and supports connections to either supplied or free issue equipment is included. Fittings and the like include, bends with turning vanes, tees equal and reducing, blank ends, transformations, flexible connections, diffuser and grille connections, stiffeners, branch spigots, volume control dampers, standard operated multi-leaf fire dampers, insulated access doors, flanges, tie rods, supports and brackets fixed to masonry construction. Also included is the pressure testing and cleaning of internal surface of ductwork, temporarily disconnection and reconnection of grilles and diffusers, setting to work and regulation of air system.

Ductwork Standard - HVCA DW 142

Circular low velocity galvanized mild steel ductwork (straight seams). t 6,500

Circular high velocity galvanized mild steel ductwork (straight seams). t 7,500

PRINCIPAL RATES

		£
Rectangular low velocity galvanized mild steel ductwork	t	8,000
Rectangular high velocity galvanized mild steel ductwork	t	9,500

AIR HANDLING UNITS

Packaged air handling unit comprising of:
Fresh/recirculated air mixing box section
Fresh/exhaust/recirculated air dampers
Bag filter assembly
Heating coil section
Centrifugal fan section
Installed in a double skin, modular framed enclosure and complete with access section, anti-vibration mountings and flexible connections. Unit suitable for external location

Air Volume vs External Resistance

2700 m3/hr vs 150 pa	nr	4,200
3600 m3/hr vs 250 pa	nr	4,800
7200 m3/hr vs 250 pa	nr	6,200
28800 m3/hr vs 500 pa	nr	16,800

Packaged air handling unit comprising of:
Fresh/recirculated air mixing box section
Fresh/exhaust/recirculated air dampers
Bag filter assembly
Main heating coil section
Cooling coil section
Re-heat section
Electric operation steam humidifier
Centrifugal fan section installed on a double skin, modular framed enclosure and complete with access sections, anti-vibration mountings, and flexible connections. Unit suitable for external location

Air Volume vs External Resistance

2700 m3/hr vs 150 pa	nr	9,230
3600 m3/hr vs 250 pa	nr	10,200
7200 m3/hr vs 250 pa	nr	13,300
28800 m3/hr vs 500 pa	nr	32,520

MECHANICAL AND ELECTRICAL WORK

AXIAL FANS

Aerofoil axial flow fan, long casing including supports, anti-vibration mountings, standard flexible connections and matching flanges.

Fan Diameter	Air Volume	External Resistance		£
300 mm	1800 m3/hr	150 Pa	nr	400
300 mm	1800 m3/hr	250 Pa	nr	420
475 mm	3600 m3/hr	150 Pa	nr	475
475 mm	3600 m3/hr	250 Pa	nr	500
610 mm	5400 m3/hr	150 Pa	nr	610
610 mm	5400 m3/hr	250 Pa	nr	640
760 mm	7200 m3/hr	150 Pa	nr	720
760 mm	7200 m3/hr	250 Pa	nr	780

Bi-furcated axial flow fan, for operating temperatures up to 200°C, including supports, anti-vibration mountings, high temperatures flexible connections and matching flanges.

Fan Diameter	Air Volume	External Resistance		
150 mm	1080 m3/hr	80 Pa	nr	350
200 mm	1440 m3/hr	100 Pa	nr	390
300 mm	2160 m3/hr	125 Pa	nr	470
375 mm	2880 m3/hr	150 Pa	nr	510

AIR INLET AND EXHAUST LOUVRES

Anodized aluminium inlet and exhaust louvre complete with 100mm deep blades, flanged frame and bird proof mesh screen. Louvres fixed to timber frame provided by others.   m2   180

Extra for powder coating paint finish   m2   15

FIRE FIGHTING EQUIPMENT

Wet and dry risers

100mm rising main including horizontal breeching inlet with 2 nr 64mm instantaneous couplings, inlet box, landing valves and boxes.

Cost per landing   nr   850

PRINCIPAL RATES

Hose reels

£

Supply pipework is measured elsewhere.

Swing type automatic recessed hose reel c/w with housing for building into masonary structure, 24 metres of hosing, isolation and drain valve.   nr   620

Fire hydrants

Supply pipework is measured elsewhere.

100mm pillar hydrant, cast iron sluice valve, flanged connections   nr   870

Underground hydrants to BS750 complete with sluice valve, flanged connections.   nr   300

Sprinkler installation

The number of sprinkler heads required to provide the necessary degree of protection is determined by the hazard classification. The design of sprinkler installation is carried out by registered companies who are on the Fire Office Committee (FOC) list.

As a guide either the cost per m2 of protected area or cost per head can be used. In both cases the rates cover for all associated pipework, valve sets, booster pumps and supports.

Cost for area protected   m2   £9.00-16.80

Cost per sprinkler head   nr   £147.00

(Storage and provision of standby power equipment is not included.)

CALORIFIERS

Copper storage calorifiers

Calorifiers to BS853 either horizontal or vertical pattern complete with cradle or legs, all pipe connections, mountings and instruments, thermostatic controls for primary side, temperature and pressure relief valves, mineral fibre thermal insulation and hammer clad aluminium finish. Heater battery fed with steam or LTHW as listed sized to raise stored water from $10^{\circ}C$ to $65^{\circ}C$ in one hour.

MECHANICAL AND ELECTRICAL WORK

Heating Source - Steam (L.P.) £

Capacity

500 litres	nr	1200
1000 litres	nr	1635
2000 litres	nr	2315
3000 litres	nr	2910

Heating Source - LTHW

Capacity

500 litres	nr	1475
1000 litres	nr	1855
2000 litres	nr	3100
3000 litres	nr	3985

Non-storage calorifiers

Steam fed non-storage calorifier providing LTHW service $82°C/71°C$, horizontal pattern c/w with supports, steam traps, sight glasses, valves, mineral fibre thermal insulation and hammer clad aluminium finish, temperature controls.

Heating Capacity

100 kw	nr	554
200 kw	nr	700
300 kw	nr	1065
600 kw	nr	1970
1000 kw	nr	3200

PRINCIPAL RATES

ELECTRICAL WORK

CONDUIT SYSTEMS

		Run on Surface £ p	Run in Chase £ p
The following rates include for installing materials up to a maximum height of 4.5m, delivery, off loading storage, distribution up to room, wastage.			
Galvanized heavy gauge steel conduit to BS4568 Class 4 inclusive of elbows, tees, circular steel boxes, intersections, saddles, clips and draw wires. Cable measured separately. Cable diameter			
16 mm	m	30.38	29.12
20 mm	m	32.62	32.27
25 mm	m	42.70	39.55
32 mm	m	51.73	49.49
Black enamel, heavy gauge, steel conduit to BS4568 class 2 inclusive of elbows, tees, circular steel boxes, intersections, saddles, clips and draw wires. Cable diameter			
16 mm	m	24.92	23.66
20 mm	m	27.16	26.74
25 mm	m	35.42	32.27
32 mm	m	44.38	42.14
PVC conduit, super high impact, heavy gauge, inclusive of elbows, tees, circular steel boxes, intersections, saddles, clips and draw wires. Cable diameter			
16 mm	m	20.16	18.99
20 mm	m	23.94	22.23
25 mm	m	31.23	29.43
32 mm	m	44.28	43.02

MECHANICAL AND ELECTRICAL WORK

TRUNKING SYSTEMS

Steel trunking single compartment, heavy gauge galvanized to BS4678, inclusive of full length lids, fillets, gusset elbows, tees, fourway section, blank ends, sleeve complings, reducers, cable retainers, fire barriers and supports.
Trunking size

		£ p
50 X 50 mm	m	46.69
75 x 50 mm	m	52.50
75 x 75 mm	m	59.01
100 x 50 mm	m	60.62
100 x 75 mm	m	64.61
100 x 100 mm	m	70.70
150 x 150 mm	m	85.40
150 x 75 mm	m	91.98
150 x 100 mm	m	99.12
150 x 150 mm	m	107.80
225 x 100 mm	m	145.67

PVC trunking, single compartment to BS4678 and 4607, inclusive of full length lids, couplings, stop ends, flat angles, internal/external angles, flat tees, adaptors, side tees, intersections, circular boxes and supports.

mini-trunking

16 x 16 mm	m	9.84
25 x 16 mm	m	11.40
32 x 12.5 mm	m	11.40
40 x 16 mm	m	13.44
40 x 25 mm	m	15.68
40 x 40 mm	m	20.32
50 x 25 mm	m	19.84
50 x 32 mm	m	21.88

PRINCIPAL RATES

standard trunking

		£ p
50 x 50 mm	m	50.40
75 x 50 mm	m	56.88
75 x 75 mm	m	70.20
100 x 50 mm	m	79.62
100 x 75 mm	m	88.32
100 x 100 mm	m	101.22
150 x 75 mm	m	166.74
150 x 100 mm	m	206.88
150 x 150 mm	m	216.24

CABLES

Multicore armoured PVC insulated cables to BS6346 inclusive of all clips, fixings, cable terminations. Installed on surface, tray or trunking.

			Cost £/m	
		2 core	3 core	4 core
1.5 mm2	m	3.57	4.05	4.41
2.5 mm2	m	4.16	4.77	5.33
4 mm2	m	6.17	7.29	9.09
6 mm2	m	7.67	9.51	11.04
10 mm2	m	11.30	14.42	14.75
16 mm2	m	11.87	18.12	18.69

Single core PVC insulated cables, non-armoured with sheath (twin and earth) BS6004, inclusive of all clips, fixings, installed on surface or in chases.

		£ p
1.0 mm2	m	1.91
1.5 mm2	m	2.16
2.5 mm2	m	2.57
4 mm2	m	4.41
6 mm2	m	5.99
10 mm2	m	9.60
16 mm2	m	14.09

MECHANICAL AND ELECTRICAL WORK

TRAY AND LADDER SYSTEMS £ p

Cable tray, standard pattern heavy duty from perforated mild steel to BS1449, hot dip galvanized, inclusive of flat bends, outside/inside risers, tees, 4-way cross pieces, reducers.

tray width

102 mm, gauge 1.5 mm	m	39.12
152 mm, gauge 1.5 mm	m	44.94
229 mm, gauge 1.5 mm	m	57.72
305 mm, gauge 1.5 mm	m	72.78
457 mm, gauge 2.0 mm	m	110.16
610 mm, gauge 2.0 mm	m	142.62

Cable tray, return flange, heavy duty formed from perforated mild steel to BS1449, hot dip galvanized, inclusive of flat bends, outside/inside risers, tees, 4-way cross pieces, reducers.

tray width

102 mm, gauge 1.5 mm	m	62.85
152 mm, gauge 1.5 mm	m	72.45
229 mm, gauge 1.5 mm	m	82.65
305 mm, gauge 1.5 mm	m	102.98
457 mm, gauge 2.0 mm	m	149.03
610 mm, gauge 2.0 mm	m	185.40

Cable ladders heavy duty, 130mm deep, 2 mm gauge, formed from mild steel to BS1449, galvanized to BS729 inclusive of internal/external riser, flat bends, equal tees, four way cross pieces, reducers, couplers, cantilever brackets, fixings and supports.

ladder width

152 mm	m	115.38
302 mm	m	130.56
452 mm	m	146.88
602 mm	m	161.76
752 mm	m	180.84
902 mm	m	192.60

PRINCIPAL RATES

FIRE ALARM SYSTEMS

£ p

System and installation to BS5839 'Fire Detection and Alarm Systems in Buildings', Local Fire Officer but excluding the requirements of the Employers' Insurers. Each zone and sounder circuit is monitored. Battery back up for 72 hours provided. Fire Alarm panel to BS3116 part 4. Wiring in red over sheath MICV cable inclusive of all clips, fastenings and terminations. Rates inclusive of system commissioning by the fire alarm equipment manufactuer.

break glass	nr	102.00
bell	nr	114.00
heat detector	nr	126.00
smoke detector	nr	126.00

Extra for Zone Panel, battery charger

4 zone	nr	760.00
8 zone	nr	1275.00
16 zone	nr	1736.00
Extra for repeater panel	nr	400.00

LIGHTNING PROTECTION

Lightning protection system and installation to BS6651 "Protection of Structures against Lightning". System comprises of air termination, down conductors, earth rods, clips and fastenings, junction clamps, disconnection links, rod to cable lugs and sockets.

Total area of building protected

750 m2	nr	5,500
1600 m2	nr	6,800
3600 m2	nr	8,700

The rates are to be adjusted for complex roof profiles.

MECHANICAL AND ELECTRICAL WORK

POWER AND LIGHTING INSTALLATIONS £ p

Power

Rates for the wiring of power points including switched socket outlets, fused spurs, consumer unit, fixings, conduit system and associated components as applicable. Accessories used are of high impact plastic manufacture.

Power point PVC twin and earth cable in residential property on ring main with MCB circuit protection	nr	42.00
PVC insulated cables on galvanized conduit system in commercial property	nr	74.00
PVC insulated cables in galvanized conduit system in industrial property	nr	87.00

Extra for

110v industrial circuit transformer (30VA)	nr	25.00
110v socket complete with isolator	nr	60.72
415v 16 amp socket complete with RCD protection (30mA)	nr	141.84
45 amp cooker control unit	nr	11.60
shaver socket	nr	27.34
1.7 kw storage heater	nr	160.53
2.5 kw storage heater	nr	210.87
3.4 kw storage heater	nr	243.30

Busbar installation, maximum rating 200 amp, inclusive of end connection isolator units, supports and fixings to masonry construction.

length 15 m	m	29.00
length 30 m	m	46.50
length 60 m	m	81.00
Extra over for fused tap-in box	nr	50.00
Extra over for isolator fused tap-in box	nr	94.00

Lighting

Rates for the wiring of lighting points including light switch, connection unit and accessories but excluding lighting fittings. Wiring from consumer unit or distribution board.

PRINCIPAL RATES

Lighting Point   £ p

PVC twin and earth cable in residential property, chased into plaster, covered in PVC sheath	nr	40.00
PVC insulated cables in galvanized conduit system in commercial property	nr	90.00
PVC insulated cables in galvanized conduit system in industrial property	nr	105.00

Extra for luminaires as listed inclusive of supports, flexible cable, connector blocks, control gear, conduit boxes, fixed to a masonry construction. Lamp and opal diffuser included in price where applicable.

Fluorescent tubes (standard)

1 x 18 W, length 600 mm	nr	20.23
2 x 18 W, length 600 mm	nr	32.56
1 x 36 W, length 1200 mm	nr	26.78
2 x 36 W, length 1200 mm	nr	46.49
1 x 58 W, length 1500 mm	nr	32.20
2 x 58 W, length 1500 mm	nr	57.31
1 x 70 W, length 1800 mm	nr	38.08
2 x 70 W, length 1800 mm	nr	62.61
1 x 100 W, length 2400 mm	nr	49.56
2 x 100 W, length 2400 mm	nr	80.00

Fluorescent tubes (emergency)

1 x 36 W, length 1200 mm	nr	168.20
2 x 36 W, length 1200 mm	nr	217.89
1 x 58 W, length 1500 mm	nr	136.99
2 x 58 W, length 1500 mm	nr	158.23
1 x 70 W, length 1800 mm	nr	177.83
2 x 70 W, length 1800 mm	nr	197.64
1 x 100 W, length 2400 mm	nr	207.08
2 x 100 W, length 2400 mm	nr	231.68

MECHANICAL AND ELECTRICAL WORK

Down Lights £ p

Range of prices cover for lamp type, luminaire type, finish, fixings, adjustable, recessed, wall or ceiling mountings

Luminaire	nr	£15-20

Industrial Lighting

high bay luminaires

150 W SON lamp	nr	212.00
250 W SON lamp	nr	229.00
400 W SON lamp	nr	254.00

low bay luminaires

150 W SON/T lamp	nr	169.00
250 W SON/T lamp	nr	187.00
400 W SON/T lamp	nr	206.00

heavy duty bulkhead luminaires

80 w HPL-N lamps	nr	142.50
80 w HPL-N lamp (Zone 2)	nr	152.50

flood lights

SOX - E66 lamp	nr	244.00
400 w SON/T lamp	nr	284.00
400 w HPI/T lamp	nr	329.00

INCOMING MAINS AND SWITCH GEAR INSTALLATION

Internally located substation equipment, comprising of HV switchgear, 1 No 1000 DVA transformer complete with Bukholtz relay, link disconnection unit, LV switch board, interconnecting cabling, electronic fault detection relays, inter-trip relays, cabling, break glasses, 30v dc supply, delivery, installation, works and site testing and commissioning.

    Approximate cost   £70,000

Main building, switch board panel, of modular construction. Maximum rating 800 amps, inclusive of delivery, off loading, site assembly, making all final cable connections, instrumentation and testing.

    Approximate cost   £10,000

The above figures can vary substantially depending on the complexity and requirements of the installation, location of site and local electricity authority requirements and the necessary adjustments must be made.

PRINCIPAL RATES

LIFTS AND ESCALATORS INSTALLATION £

Lifts

The following rates assume a floor to floor height of 3 metres with standard finish to cars and doors.

Passenger lifts

Electrically operated, 8/10 person lift serving 6 levels and travel speed of 1 metre/second.	nr	49,000
Extra for bottom motor room	nr	2,800
Extra levels served	nr	2,400
Increased speed of travel		
1 to 1.6 metre/second	nr	2,400
1 to 2.5 metre/second	nr	30,000
Enhanced finish to car	nr	6,500
Electrically operated 21 person lift serving 6 levels and travel speed of 1 metre/second.	nr	61,250
Extra for bottom motor room	nr	3,400
Extra levels served	nr	2,600
Increased speed of travel		
1.6 to 2.5 metre/second	nr	29,000
Enhanced finish to car	nr	6,500

Goods lifts

Electrically operated, two speed, general purpose lift serving 5 levels, maximum load 500 kg, manually operated shutters and automatic push button control, with a travel speed of 0.25 m/s	nr	31,800
Extra levels served	nr	2,300
Increased capacity to 2000 kg	nr	12,000
Increased travel speed 0.25 to 0.5 metre/second	nr	1,900

MECHANICAL AND ELECTRICAL WORK

		£
Electrically operated heavy duty goods lift serving 5 levels, maximum load 1500 kg, with manually operated doors and automatic push button control and speed of 0.25 metre/second.	nr	36,700
Increased capacity to 3000 kg	nr	17,000
Extra levels served	nr	2,300
Through car	nr	3,000
Oil hydraulic operated, heavy duty goods lift serving 4 levels, to take 500 kg load with manually operated shutters and automatic push button control and speed of 0.25 metre/second.	nr	41,700
Increased capacity to 1000 kg	nr	2,300
Increased capacity to 1500 kg	nr	4,100
Increased capacity to 2000 kg	nr	8,200

Escalator Installations

30° pitch escalator with a rise of 3.5 metres and enamelled sheet steel or glass balustrades.

600 mm tread	nr	34,300
800 mm tread	nr	36,750
1000 mm tread	nr	39,200
Extra for		
stainless steel balustrades	nr	1,250
glass balustrades with lighting	nr	1,250

PART FOUR

Reclamation, Landscaping and Environmental Work

9 Principal rates
10 Composite rates

9
Principal rates

The most commonly used items in soft and hard landscaping and environmental works are presented in this chapter together with the appropriate rates.

Soft landscaping

		£ p
Topsoil preservation		
Lift turf by hand for preservation to stack 100m distance	m2	3.00
Excavating topsoil for preservation by hand		
average depth 100mm	m2	1.60
average depth 200mm	m2	3.00
average depth 300mm	m2	4.50
Excavating topsoil for preservation by machine		
average depth 100mm	m2	0.40
average depth 200mm	m2	0.75
average depth 300mm	m2	1.00
Excavating		
To reduce levels by machine		
maximum depth not exceeding 0.25 m	m3	2.80
maximum depth not exceeding 0.50 m	m3	2.60
maximum depth not exceeding 1.00 m	m3	3.50
maximum depth not exceeding 2.00 m	m3	4.00
maximum depth not exceeding 3.00 m	m3	6.00
Trenches for hedges and back filling with selected soil, dispose of surplus by hand	m	3.00

RECLAMATION, LANDSCAPING, ENVIRONMENTAL

		£ p
Pits for trees and back filling with selected soil, dispose of surplus by hand		
small	nr	5.00
medium	nr	10.00
large	nr	20.00
Disposal by machine		
Excavated material off site a distance of 1 Km	m3	8.00
Excavated material off site a distance of 5 Km	m3	10.00
Excavated material on site a distance of 100m	m3	3.50
Excavated material on site a distance of 300m	m3	5.00
Filling to make up levels by hand		
obtained from on site spoil heaps 100 m distance	m3	21.00
obtained off site to specification	m3	42.00
Filling to make up levels by machine		
obtained from on site spoil heaps 100 m distance	m3	10.00
obtained off site to specification	m3	31.50
Filling to planters on site		
obtained from on site spoil heaps	m3	45.00
Surface treatments		
Applying herbicides to specification	m2	0.35
Preparing sub soil for top soil with plough or cultivation	100m2	6.25
Rolling surface with light roller for grass areas	100m2	6.75
Hard landscaping		
Paving slabs bedded in lime mortar, pointed		
precast concrete natural finish on sand and hardcore		
600 x 450 mm, 50mm thick	m2	11.50
600 x 600 mm, 50mm thick	m2	10.75
600 x 750 mm, 50mm thick	m2	10.35
600 x 900 mm, 50mm thick	m2	10.00

PRINCIPAL RATES

		£ p
Precast concrete natural finish on sand and hardcore		
600 x 450 mm, 63mm thick	m2	14.25
600 x 600 mm, 63mm thick	m2	13.20
600 x 750 mm, 63mm thick	m2	12.70
600 x 900 mm, 63mm thick	m2	12.25
Marshall's 'Keyblok' compacted on sand, vibrated		
200 x 100 mm paving blocks, 65mm thick		
straight joints, half bond	m2	14.50
herringbone pattern	m2	15.00
Marshall's 'Keyblok' compacted on sand, vibrated		
200 x 100 mm paving blocks, 80mm thick		
straight joints, half bond	m2	16.00
herringbone pattern	m2	16.75
Marshall's 'Monolok' compacted on sand: vibrated		
190 x 105 mm interlocking paving blocks, 60mm thick	m2	16.00
190 x 105 mm interlocking paving blocks, 80mm thick	m2	17.50
Brick paving on concrete base bedded jointed and pointed in cement mortar		
staffordshire blue paviors, 50mm thick	m2	51.00
rough stocks, 65mm thick	m2	30.00
Granite sett paving on concrete base bedded jointed and grouted in cement mortar		
new setts 100mm thick	m2	42.50
new setts 150mm thick	m2	55.00
reclaimed 100mm thick	m2	38.00
reclaimed 150mm thick	m2	48.00
Macadam and asphalt pavings laid on prepared bases (up to 500m2)		
Coated macadam basecourse		
40mm open textured material, 60mm thick	m2	4.25
40mm open textured material, 80mm thick	m2	5 10

RECLAMATION, LANDSCAPING, ENVIRONMENTAL

		£ p
40mm dense material 60mm thick	m2	5.10
40mm dense material 80mm thick	m2	6.10
Coated macadam wearing course		
10mm open textured material 20mm thick	m2	2.70
10mm open textured material 25mm thick	m2	3.00
10mm dense material 20mm thick	m2	3.00
10mm dense material 25mm thick	m2	3.30
fine cold asphalt wearing course 15mm thick	m2	2.70
fine cold asphalt wearing course 25mm thick	m2	3.50
Concrete roads and paths laid on prepared base		
150mm - 300mm thick	m3	60.00
Formwork to edges of beds, 150-300mm	m	2.50
Fabric reinforcement in concrete roads per layer		
A142	m2	2.00
A193	m2	2.35
Joints with fibreboard and sealant		
150mm deep	m	3.50
300mm deep	m	5.50

Special surfacing

		Black	Coloured
Playtop '85' impact absorbing safety surfacing laid on existing surfaces from Charles Lawrence UK Ltd		£ p	£ p
25 mm to falls, cross falls	m2	24.75	46.50
60 mm to falls, cross falls	m2	47.70	67.25
80 mm to falls, cross falls	m2	53.00	77.50
100 mm to falls, cross falls	m2	57.00	81.50

Fire paths

		£ p
Grass concrete paving using polystyrene formers and grass seed		
100mm thick	m2	14.50
150mm thick	m2	17.50

PRINCIPAL RATES

		£
Grass block paving using concrete units and grass seeds		
103 mm thick units	m2	18
Street furniture		
bollards		
plain concrete	nr	50
exposed aggregate concrete	nr	85
cast iron plain	nr	100
cast iron ornamental heavy duty	nr	200
Extra for excavating holes and concrete foundations	nr	15
Litter bins		
plain concrete	nr	117
reinforced concrete	nr	140
Benches and seats		
Boulevard quality	nr	210
standard quality	nr	120
extra for excavating holes and concrete foundations	nr	10
Bicycle stands		
concrete block in paving	nr	14

ENVIRONMENTAL WORKS

Demolition

Demolishing down to ground level and disposal of materials		
brick dwarf walls	m	8
timber fences	m	5
chain link fences	m	4
brick garages (single)	nr	850
concrete steps	nr	25
Take up and dispose of materials		
concrete paths	m2	2
tarmac paths	m2	1

RECLAMATION, LANDSCAPING, ENVIRONMENTAL

		£ p
Cutting down trees and grubbing up roots, remove from site		
small	nr	20.00
large	nr	135.00
Excavating		
bulk filling to make up levels, imported material	m3	12.00
cultivating surfaces of natural ground 150mm deep	m2	0.25
removing turf, cultivating ground 150mm deep importing soil, dressing and applying bark mulch	m2	1.75
seeding prepared ground, rolling and first cut	m2	0.30
imported turf on prepared ground and first cut	m2	3.50
imported top soil with soil improver additive	m3	11.00

Fencing, railings and gates

Strained wire with concrete posts		
1.0 m high with 6 wires	m	10.00
extra for end post	nr	50.00
extra for corner post	nr	60.00
Strained wire with wood posts		
1.5 m high with 8 wires	m	6.00
extra for end post	nr	20.00
extra for corner post	nr	21.00
Chain link fencing with galvanized mesh with galvanized steel angle posts encased in concrete foundation including excavation		
1800 mm high	m	15.00
extra for end post	nr	33.00
extra for corner post	nr	55.00

PRINCIPAL RATES

		£
Chain link fencing with plastic coated mesh with concrete posts encased in concrete foundation including excavation		
1800 mm high	m	17
extra for end post	nr	55
extra for corner post	nr	70
Pair of gates and gate posts with plastic chain link fencing with hinges, locking bar, lock, bolts, gate posts, strutted in concrete foundation including excavation	nr	450
Wooden fencing close board with oak posts two 76 x 38mm arris rails in mortices 100 x 16mm sawn feather edge vertical boards one side all pressure impregnated and painted with Sadolin; concrete back filling around posts;		
900 mm high	m	20
1800 mm high with 3 arris rails and gravel board	m	32
Gates; wrought softwood, framed ledged and braced, with oak posts including all ironmongery, all pressure impregnated and painted with Sadolin, concrete back filling around posts		
850 x 900mm high	nr	85
850 x 1800mm high	nr	110
Railings in galvanized mild steel		
20 mm diameter railings at 130 mm centres in panels 3 m long, between 50 x 15 mm rails top and bottom set screwed to lugs welded on 100 diameter RHS posts with cast iron capping, base set in concrete foundation 700mm deep including excavation		
1200mm high	m	46
1500mm high	m	60
extra for curved	m	10
extra for sloping ground	m	4

10 Composite rates

SOFT LANDSCAPING

Site Clearance

		£
Clear vegetation and small trees, to tip not exceeding 10 km distance	m2	0.10-0.40
Cut down tree, remove stumps to tip not exceeding 10 km distance		
500-1000mm girth	nr	17-20
1-2 m girth	nr	33-36
2-3 m girth	nr	130-135
3-5 m girth	nr	650-750
exceeding 5m girth	nr	850-1000
Extra for filling holes with excavated material	nr	5-10
concrete	nr	20-100
Lift turf and stack on site by machine	m2	0.50-1.50
Demolish buildings to ground level remove material from site		
brickwork	m3	0.60-0.75
reinforced concrete	m3	2.50-3.00
steel frame with cladding	m3	0.45-0.65
timber framed with timber cladding	m3	0.20-0.25

RECLAMATION, LANDSCAPING, ENVIRONMENTAL

Earth moving

The cost varies considerably depending on the quantity to be moved and the size and type of machine available.

Assuming the spoil is not to be taken off site the following shows comparative costs between using Bulldozers and Motorized scraper per cubic metre.

	length of haul (one way)		
Crawler Dozer	50m	100m	150m
	£ p	£ p	£ p
D5 90 kw m3	2.50	4.75	6.95
D6 90-125 kw m3	2.25	4.25	6.50
D7 - D8 125-250 kw m3	2.15	4.10	6.25
	1000m	2000m	3000m
	£ p	£ p	£ p
Motorised scrapers (621E) 15.3 m3	2.40	2.93	3.96
(631E) 23.7 m3	2.17	2.75	3.71
(651E) 33.6 m3	2.06	2.50	3.45

Rates for general reduce level excavation, spread and level on site not exceeding 50 m distance.

Depth		
0.25m	m3	6.30
1.00m	m3	7.00
2.00m	m3	7.80
3.00m	m3	9.70

If double handling is necessary these rates could double. Removal of artificial obstructions such as buried foundations would increase the rate by the volume of material removed. The cost would range between £10-30 m3 extra depending on the degree of difficulty encountered.

COMPOSITE RATES

Groundwork

		£
Excavate top soil from deposits in spoil heaps, after treatment with weed killer, excavate and spread in layers over site and grade to finished levels, minimum 300 mm thick by machine	m3	10-12
extra for importing top soil	m3	3-5
extra for hand working over surface	m2	2-3
Grass seeding after preparation of soil, including applying fertilizer, raking, weeding rolling and after treatment	100 m2	25-35
Laying turf on previously prepared ground, apply fertilizer and pesticide lightly roll and cut with machine after period of growth	m2	3-4
Hydraulic mulch seeding by Specialists (Comtek UK) on previously prepared ground on areas over 5 ha		
with high clay content sub-soil	ha	2000-3000
with no fines sub-soil (rock, sand)	ha	3000-4000
Maintenance of grass recreation areas for first year including mowing, fertilising, watering, repairing	ha	2750-3000
Artificial grass, polythene pile	m2	25-30
Polypropylene yarn on latex backing	m2	10-15

Planting

Excavate pit, break up bottom incorporate enriched soil stake and tie back fill level ground		
trees PC £30	nr	40-45
trees PC £15	nr	20-25
trees PC £5	nr	10-15
Mature trees 10m high including excavation, braced supports, tree guard and grid	nr	500-600

RECLAMATION, LANDSCAPING, ENVIRONMENTAL

		£
Hawthorn hedge, double row at 300 mm centres in trench back fill with enriched soil	m	6-10
Beech hedge, double row at 450 mm centres in trench back fill with enriched soil	m	10-15
Privet hedge, single row at 600 mm centres in trench back fill with enriched soil	m	5-7
Various shrubs at PC £1.50 - 3.00	nr	2-4
Various plants at PC £1.00 - 3.00 each	m2	10-50

HARD LANDSCAPING

Street furniture

Bollards including excavation and concrete foundation		
precast exposed aggregate concrete	nr	85-110
steel	nr	150-175
cast iron	nr	150-200
Litter bins in precast concrete with galvanised liner	nr	110-150
Precast concrete block bicycle stands set in paving	nr	12-15
Bench seats in hardwood with precast concrete supports set into paving	nr	100-250
Plant containers in precast concrete	nr	100-250

Paving

Tarmacadam roads two layers 75mm thick including excavation and sub-base	m2	15-25
Concrete roads 150mm thick including excavation and sub-base	m2	25-35
Precast concrete paving slabs 50 mm thick on hardcore and sand including excavation	m2	12-17
York stone paving slabs including excavation	m2	50-60
Patent block type paving in sand on hardacre and sand including excavation	m2	20-30

COMPOSITE RATES

		£
Brick paving in Staffordshire Blue or Accrington Red clay bricks in cement mortar including excavation	m2	45-55
Granite setts in cement mortar including excavation	m2	55-65
'Grass block' patent paving with sand blinding, soil filling, grass seeding including excavation	m2	15-20
Fire path access paving blocks over grass including excavation	m2	20-30

Kerbs and edgings

		£
50 x 150 mm precast concrete edging with concrete foundation	m	5-7
125 x 250 mm precast concrete kerbs with concrete foundation and haunching	m	7-10
125 x 250 mm precast concrete channels to last	m	7-10
Precast concrete drainage channels with concrete foundation	m	20-30
extra for silt box complete	nr	100-110
Polyester concrete surface water drainage channels with galvanized gratings	m	50-100
extra for sump units and trap	nr	50-60

ENVIRONMENTAL WORKS

Fences and gates

Softwood impregnated and rail	m	15-20
extra for oak posts and rails	m	5-7
Close boarded softwood impregnated with softwood posts 1.6m high	m	25-30
extra for concrete posts	m	7-10
Chain link galvanized mesh 1.6m high	m	15-20
extra for plastic coated mesh	m	2-3
extra for pair of gates 2.5m wide	nr	350-450
extra for three rows of barbed wire	m	3-4

RECLAMATION, LANDSCAPING, ENVIRONMENTAL

		£
Chain link fencing for tennis court including gate 2.75m high	nr	3500-4000
Tennis court surfacing in macadam (one court)	nr	8000-10000
extra for tournament quality surfacing	nr	10000-25000
Security galvanized steel palisade fencing to high standard specification 3.00 m high	m	35-45
extra for gates	pair	750-1000
Anti-intruder chain link fencing with concrete posts cranked at top with 3 lines of barbed wire 2.90m high	m	55-60
extra for gates	pair	900-1500

Playground equipment from SMP (Playgrounds) Ltd, Chertsey.

Climbing equipment		Cost of supply	Cost fixed £
Novelty shapes single person use	PC	£125-250	175-200
Large frames with slides, multi-person use	PC	£1500-4000	2000-4500
Towers and links between for Seniors			
towers	PC	£500-800)	
bridges, agility links between	PC	£750-1500)	2000-3000

Swings

Multi-seat swings			
2 seat	PC	£550-600	650-700
4 seat	PC	£1000-1100	1150-1250
Extra for 'special need' seats	PC	£25	
Multi-person swingabout	PC	£1850	2000

Slides

contour slides for grass mounds	PC	£750-2000	1000-2500
junior slides for level ground	PC	£750-1500	1000-1600
Senior Action Pack with selection of towers, links and access components	PC	£2150-24500	2500-30000

These can be supplied to individual requirements with an infinite variety of units

COMPOSITE RATES

Goal posts	Cost of supply			Cost fixed £
senior size (per set of two cross bars and four uprights)	PC	£390	)	
extra for sockets per upright	PC	£25	)	
extra for net irons per goal	PC	£40	)	500
Picnic tables made from oiled hardwood with seats each side	PC	£300–400		350–450
Wooden animals (from ponies to elephants)	PC	£350–930	Nr	400–1000
Playscreed surfacing				
20mm thick tiles 750 x 750	PC 38m2		m2	55–65
50mm thick tiles 750 x 750	PC 58m2		m2	75–85

Playground equipment from Wicksteed Leisure Ltd, Kettering

In addition to similarly priced climbing equipment, swings and slides, various other products available include:

Roundabouts	PC	£700–2000	1000–280
See-saw type units	PC	£400–1100	500–1200
Units for nursery schools	PC	£150–750	200–800
Fun ball games	PC	£600	700
Multiplay indestructible polyethylene cubes and tunnel units	PC	£550–2000	600–2500

PART FIVE

Refurbishment

11 Composite rates

11
Composite rates

BUILDING WORK

The wide variation in building construction of buildings requiring refurbishment preclude giving a comprehensive list of rates but for a budget estimate the following will be of assistance where full details are not known.

DEMOLITION

Demolition of entire structures are costed by the volume of the building, the following are a guide for small buildings, for large buildings competitive quotations should be obtained because the value of the salvaged material can have a considerable effect on the price charged.

			£
Single storey brick buildings	50-100 m3	m3	6-9
Single storey brick buildings	100-200 m3	m3	5-6
Two storey brick buildings	200-400 m3	m3	2-5

Alterations internally

Cutting openings size 1.5-2.0 m2 through walls	typical	2-5
150mm blockwork	nr	250-350
225mm brickwork	nr	375-500
Taking down walls		
partition, 100-150mm blockwork	m2	7-10
brickwork		
half brick thick	m2	10-15
one brick thick	m2	15-20
Fill openings in walls		
partition, 100-150mm blockwork	m2	20-25
brickwork		
half brick thick	m2	35-40
one brick thick	m2	60-70

REFURBISHMENT

		£
Taking out fireplaces and fill openings only	nr	50-100
Cutting holes through floors for stair 2 - 3 m2 in reinforced concrete, thickness		
150 mm	nr	450-500
225 mm	nr	650-1000
Take out timber stair and fill in timber floor opening	nr	150-250
Taking out bathroom fittings per bathroom	nr	150-200
Taking out kitchen fittings per kitchen	nr	100-150
Taking up floor coverings and making good surface to receive new covering		
timber including removing nails	m2	3-5
screed including making good	m2	5-7
Taking up skirtings, dado rails	m	1-2
Pull down ceilings prepare to receive new ceiling finishes	m2	2-5
Hack off wall finishings	m2	2-5

Alterations externally

Take off roof coverings		
tile and dispose	m2	3-4
slates and set aside for reuse	m2	4-6
asphalt roofing and skirtings	m2	7-9
Extra for asbestos or other toxic waste products	m2	25-30
Take down gutters and fascia boards	m	2-5
Take down chimney stack to below roof slope	nr	300-350
Take down chimney breast from roof to ground floor	nr	1000-1250
Forming large opening in walls (up to 5m2)	nr	250-350
Allowance for erecting screens	m2	10-20
Provision of raking shores to two storey building	nr	300-400
Provision of dead shores to first floor level	nr	200-300

COMPOSITE RATES

CIVIL ENGINEERING WORK £

Defective concrete repairs

Cutting out cracks in concrete and renovating	m	5-10
Filling openings in floors with new reinforced concrete bonded to existing		
150mm thick	m2	35-40
250mm thick	m2	45-50
Inserting additional beams under existing concrete floors (0.25-0.50 m2)	m	250-300
Inserting additional columns	m	500-750
Forming new bases in existing basement	Nr	1500-3000
Cutting through reinforced concrete		
floors 250mm thick	m2	65-75
walls 150mm thick	m2	50-60

Structural steelwork

Taking down existing steel beam encased in concrete, renewing with larger and casing in concrete	m	200-250
Erecting new column 3m high including providing new base in existing floor making good, casing in concrete and making connections	Nr	3000-4000

PART SIX

General Data

12 Life cycle costing
13 The development process
14 Professional fees
15 Construction indices
16 Property insurance

12
Life cycle costing

INTRODUCTION

Life cycle costing is a system for budgeting and controlling the costs of the design, development and property management of a building. It should focus on the building owner's policy requirement for the building.

The life cycle cost of a building is the total cost commitment to that building. It is the sum of the initial capital costs and the future running costs. These costs are incurred at different times and so they need to be discounted back to the present to allow them to be compared to the initial capital costs. There are standard techniques and published tables for discounting future costs and expressing them as a single sum of money (Net Present Value or NPV) or as an annual flow of money (Annual Equivalent).

DEFINITIONS

A range of terms are used in life cycle costing and the meaning of the most common terms is defined below.

Life Cycle Cost Planning

The systematic and objective quantification of the initial and life time costs of a building or a building element at the design stage to produce a scheme which satisfies the client's objectives.

Costs in Use

The technique of converting the life time cost consequences of a particular design decision into a single consistent measure of cost which can be used for comparison purposes. The 'measure' may be a single sum of money (NPV), a flow of money per time period (Annual Equivalent) or it may be expressed in terms of scarce resources whose use is to be minimised such as energy or manpower.

Life Cycle Cost Analysis

The systematic collection and analysis of the running costs and performance characteristics of a building in use. The analysis should reflect the degree to which the costs incurred are justifiable in terms of the performance delivered.

The purpose of the analysis is to provide appropriately structured and processed data which can be used to prepare Life Cycle Cost Plans for new buildings or the Life Cycle Cost Management of an existing building.

GENERAL DATA

Life Cycle Cost Management

The process for developing and implementing a maintenance policy which recognizes that the building owner or occupier's interest may be served best by setting maintenance decisions into his broader financial context and in a time frame which goes beyond the immediate maintenance horizon. The purpose of life cycle cost management is to establish maintenance management as a creative activity rather than one which is routine or reactive.

APPROACH

The inherent weakness of life cycle cost studies is that the formative influences on a building's life cycle characteristics are the decisions taken at the early design stages when there is little or no project specific information available which can be used to test the validity of those early decisions. Furthermore, the appraisal of design options for life cycle costing purposes must assume that cost implications can be reliably predicted for many years into the future. This is a somewhat fragile assumption given the long life of buildings and that the costs themselves are the result of a complex interplay between performance, use behaviour and changes in the economic environment.

The main emphasis for the developer or owner of the building should be to concentrate on generating an outline brief and management policy in life cycle terms. Concentrating on the outline brief has the significant benefit of defining the objectives for the completed development at a stage when they are still capable of affecting design decisions. The attention paid to formulating a management policy recognizes the fact that costs over the life of a building become less and less predictable the further they are removed from the present. The owner's interests may be served also, therefore, by adopting a system which can deal with these future costs as and when they arise in a manner which is consistent with the project's objectives, which themselves will have developed from the initial definition.

The intermediate stages between the outline brief at the commencement of design and the management policy for the building in use, should be used to translate the original requirements into a physical structure which is sufficiently robust to allow subsequent management decisions to moderate its performance in response to evolving objectives and environment conditions.

BENEFITS

The benefit of adopting a life cycle approach comes from the recognition that buildings are a long term investment. Assessing buildings' value from the initial capital costs is inconsistent with their long term nature and so attention should also be given to the future costs of buildings. Their design should result from decisions which recognize the objectives for the building over its whole life cycle.

CHECKLIST FOR LIFE CYCLE COSTING AT OUTLINE BRIEF STAGE
(used to prepare Life Cycle Cost Plan)

Define Objectives

Establish Client's interest e.g. owner occupation - develop and lease - develop and sell.
Establish Client's time horizon e.g. short term (develop) - medium term (develop and own for a limited period) - long term (develop and own long term).
Establish time horizon over which Client's objectives will remain stable.
Establish Client's specific space requirements, if for owner occupation.
Establish nature of tenants, if for leasing e.g. known tenants with known preferences - speculated tenant groups with open requirements.

LIFE CYCLE COSTING

Establish nature of lease and service charges, if tenanted e.g. - type and duration of lease - rent and service charge mix - division of responsibility between owner and tenants - basis of apportionment between tenants.
Establish importance of the project realizing a capital asset, and over what time scale this realization should take place.
Establish if there is a proposed purchaser and, if so, what his preferred requirements are.

Fit the project into the Client's financial strategy

Establish the proposed method of finance e.g. development finance - finance during holding period - long term finance.
Establish impact of project on Client's cash flow e.g. during development - during ownership - during redevelopment, refurbishment and disposal.
Establish effects on Client's capital structure e.g. gearing and other ratios - effects on future financing options - effects on market capitalization both current and future.

Fit the project into the Client's main business strategy

Establish what resources the Client can apply to the project to - develop the project - actively manage the project (who and how) - manage the completed development.

Consider the time related variables

Establish the life cycles e.g. physical life of structure and key elements - economic life related to underlying land values - functional life - social life - leasehold cycles.
Establish the coincidence of the Client's requirements with the above life cycles.
Establish time related operational constraints e.g. the criticality of the completion date.
Establish the time related marketing constraints.
Establish the time related financial constraints e.g. the running out of a line of credit.
Establish the time related contractual constraints.
Establish the time related statutory constraints e.g. changes in taxation or building regulations.
Establish time related political constraints.

Consider the finance related variables

Establish the Client's investment criteria e.g. payback - DCF - ROCE.
Establish whether the criteria measure what it is that the Client is trying to achieve.
Establish the financial objectives e.g. maximize potential for rental income - maximize potential for cash generation - minimize running cost for owner occupier - secure long term capital growth.
Establish whether the funding arrangements mean that the project must achieve a particular asset value within a specific period of time.
Establish the Client's capital and revenue constraints and how they impact on the viability of the project e.g. cost yardsticks - financial or operational criteria - allowances for enhancement or avoidance.
Establish whether the residual value of the project has a significant impact on its viability. If so, is this at the termination of the project's life - at significant intervals - annually for valuation purposes.
Establish the relationship between the value of the project and the underlying land value.

GENERAL DATA

Establish whether the ownership strategy envisages funds becoming available during the life of the project to counteract the effects of value depreciation relative to newer buildings coming on to the market.
Establish to what extent there is a trade off between capital and revenue.
Establish the Client's attitude to risk both during development and ownership.
Establish the importance of the Client's tax position regarding - the timing of cash flows - the capital cost/running cost relationship - Capital Gains Tax, or relief therefrom.
Establish whether the tax status of the prospective tenant or purchaser impacts significantly on the Client's objectives.
Establish what the tax implications surrounding the project are e.g. capital allowances - rates.
Establish whether the location of the project has fiscal implications e.g. grants - tax allowances - rates relief.
Consider the design related variables
Establish the extent of finish for the completed development e.g. shell and core - ready to decorate - ready to occupy.
Establish the quality of finish for the completed development e.g. external envelope - internal furnishings and fixtures - public and common areas - general accommodation.
Establish the degree of structural flexibility required to cater for probable or possible future developments.
Establish the degree of superficial flexibility required to cater for e.g. changes in the image of the building - changes in the function of the building.
Establish the importance of cost and time trade offs to the maintenance and refurbishment of the building in use e.g. will disruption costs be high - will rental losses be high.
Establish whether the local economy can provide for - the type of construction envisaged - the quality of fixtures and fittings envisaged - the level of maintenance envisaged.
Compile the external variables
Establish a weighted discount rate e.g. short term - long term - relative to the source of finance - relative to opportunity costs.
Establish inflation rates for e.g. project costs - operating costs - project revenues - tax position of Client or occupier.
Establish commercial risk and relate to - choice of contract - maintenance contract - best and worst scenarios of key cost elements.
Establish political risk e.g. potential changes in direct taxation and tax allowances - potential changes in indirect taxation - potential changes in local taxation - susceptibility of financial arrangements to political change.

DEVELOPMENT OF A MANAGEMENT POLICY

Once the building has been occupied information should be generated defining its true performance in use. At this stage it should be possible to establish a programme of maintenance, repair and refurbishment which accurately reflects the interplay between the Client's objectives and the performance of the building i.e. life cycle management.

LIFE CYCLE COSTING

A WORKED EXAMPLE OF A LIFE CYCLE COST PLAN

Background information

Job Title	:	New Warehouse
Location	:	North West England
Client	:	Private Development
Date	:	January 1990
Discount Rate	:	8% (cost of capital = 17%; inflation @ 8%; discount rate = 1.17/1.08 = 8%
Life Cycle	:	20 years
Job Details	:	Steel portal framed single storey building, including external works and drainage
Area	:	60,000 sq. ft.

LIFE CYCLE COST PLAN   £

1. Capital Cost

   Building; 60,000 sq ft @ £30   =   1,800,000
   External works (approx. quants)   =   300,000
   Contingencies @ 25% (Client; Design; Contract)   525,000
   Design fee @ 12.5%   325,000

                                                                    £ 2,950,000

2. Operating Costs (annual)

   2.1 General and Water rates   =   £75,000 p.a.
       Present Value (20 years)   =   735,000

   2.2 Heating and Lighting   =   £125,000 p.a.
       Present Value   =   1,225,000

   2.3 General Maintenance and Insurance =   £25,000 p.a.
       Present Value   =   245,000

                                                                    £ 2,205,000

3. Repair and Replacement Costs (intermittent)

   3.1 Redecorate every 3 years @ £2,000
       Present Value   =   5,000

   3.2 Re-carpet every 5 years @ £10,000
       Present Value   =   15,000

   3.3 Lighting and power refurbishment every 10 years @ £10,000
       Present Value   =   5,000

                                                                       £ 25,000

GENERAL DATA

LIFE CYCLE COST PLAN SUMMARY

	Summary of costs	Present Value
		£
1.	Capital Cost	2,950,000
2.	Running Cost - Operating	2,205,000
	- Repair and Replacement	25,000
	TOTAL PRESENT VALUE	£ 5,180,000

13
The development process

Attempting to explain 'the development process' in one chapter is like being asked to write a do-it-yourself guide to open heart surgery! There are excellent books covering the whole subject which are available to readers who wish to become totally involved in the subject but the following should provide an overview and an awareness to those professionals not directly involved in the development scheme. Those parties on the fringe of a project should realize the importance of their part in the success of any one development, whether it is a new-build, a refurbishment or a break-up of an existing building complex. The model chosen is a 3 acre industrial site in the North West of England for the design and build of new industrial/warehouse units. The model could just as easily have been an existing food store for break up into smaller units or a listed office building to be refurbished – the principles are the same. It may help at this stage to clarify the main players involved in any development, i.e. The Professional Team – their role and where they appear in the scheme.

The Agent

Very often the Agent/Development Surveyor will find a site and introduce it to the developer who in his opinion is the most suitable to carry out the scheme. The agent will consider the developer's own expertise in property, his track record in development, his financial status and his capability to handle the overall project by bringing together the various members of the team. The agent may be a member of the RICS/ISVA although this is certainly not essential. Individually the agent may have the necessary skills to value the site, let the completed buildings, and provide the necessary advice to secure funding for the development. Alternatively, he may bring in assistance either from within his own firm or from outside depending upon the size of the scheme and the size of the agent's firm, and also whether he is a sole practitioner or is one of the larger firms of international property consultants.

The agent therefore advises in the acquisition, the letting, the sale of the investment and will most probably provide expert input throughout the scheme. He will probably advise on the specification of the buildings and the construction of the leases to make them acceptable to the tenant and to the institutional funding market.

GENERAL DATA

The Solicitor

A solicitor will be needed at all stages to act for the developer. Firstly to acquire the site and report on the title. He will establish whether there are for example cables or sewers which have easements which can materially or adversely affect the position of buildings on the site. Clearly if a main sewer runs diagonally under the land then much of the developable area will be sterilized and therefore the land will be worth less than a "clean" site.

Secondly the solicitor will need to advise in the preparation of leases for the tenant who will occupy the factory. Generally speaking these leases will be 25 year term with 5 year upward only rent review clauses on fully repairing and insuring leases.

The solicitor will need to be fully up to date on case law and their effect on rent reviews and therefore ultimately on the value of the investment which is to be created. The lease will also need to be well drafted to take into account the good management of the estate, looking after common parts, insurance and other similar matters. It should not be so onerous that tenants refuse to sign it, but it should be fair and acceptable to the institution who may ultimately own the estate.

Lastly the solicitor will need to advise in respect of the sale of the completed development. He will act against the solicitor representing the purchaser of the scheme - most probably a pension fund.

The Architect

The agent and the developer will have their own views on trying to secure as large a building as possible on the site in order to maximize the profit. The architect may bring a sense of reality to their aspirations. He will consider the ground conditions, easements, retaining walls, rights of light, highways access, car-parking and turning areas for commercial vehicles, local planning regulations (which may differ throughout the country). He will then prepare a drawing of a workable industrial estate. He will consult the agent on what the market currently requires as well as what the investors prefer.

For example, on our imaginary 3 acre site it would be possible to build up to 60,000 sq. ft. Should the architect design for example two straight forward terraces, each of 30,000 sq. ft. facing each other across a common forecourt and easily divisible into units of say 5,000 sq. ft? Should he design say three detached buildings of varying sizes for example 10,000 sq. ft., 20,000 sq. ft., 30,000 sq. ft? Should he develop part of the site and hold some of the land back for a pre-let? After considering these matters his appraisal will then be presented to the developer and agent for discussion prior to applying for outline planning permission and building regulations.

The Quantity Surveyor

After establishing the layout, the agent and his developer client will require some costs before being able to value the site or establish the likely profit. The quantity surveyor can discuss with the agent and the architect the specification which is required, and the type of building which is proposed. Is it for the owner occupation or does it need to satisfy the funds requirements?

The developer may build, for example, to a higher specification for a fund than for an end-user. The quantity surveyor will then provide his estimate of the cost of building for the completed development not only for the building itself but also including all external works and sub-ground works. The quantity surveyor may require individual expert input in respect of engineering works, mechanical and electrical work.

THE DEVELOPMENT PROCESS

Optional Players/Reserves

It may be necessary to have in reserve some experts in particular disciplines. These may include mechanical/electrical engineers, an expert to report on ground conditions and soil surveys, a demolition expert to clear the site and grub out any foundations, a planning surveyor in the event that the planning application may end up going to appeal, a highways expert to report on any difficult traffic control or high generation of traffic, or a tax expert to advise in respect of VAT or other tax matters.

Funds Adviser

In the event that the development is to be forward funded, the fund may require their own adviser to be involved at the commencement of the development in order to monitor the quality and the progress of the scheme. In the particular model chosen it has been assumed that the development will be built speculatively and that it will be sold on completion and therefore the funds adviser does not play a part in the original professional team.

The Development

Having appointed the developer has now bought the 3 acre site and has obtained planning permission and building regulations for 60,000 sq. ft. of new industrial/warehouse units. The clock is ticking and interest is being levied in full on the site, on his acquisition fees, and any other costs he has incurred such as planning and building regulation fees.

He must now begin to move quickly and confidently to progress the scheme within the original programme. His quantity surveyors will go out to tender for the building works or will negotiate a contract with one of several known builders in order to agree the building contract within a short timescale. The developer can now start work on the site. Simultaneously, the agent will commence his marketing to get tenants for the units as quickly as possible in order to minimize the void period at the completion of the scheme.

His marketing will include preparation of brochures, erection of a signboard, a full advertising campaign in the local and national press and professional journals, as well as targeted mailshots to those companies most likely to produce a tenant. It may also be prudent towards completion of the project to hold a reception for local dignitaries, agents and the press in order to raise an awareness of the scheme. The agent should not be advising his client to accept the first tenant who wants to sign the lease because clearly the quality of the tenant, his 'covenant' is of paramount importance to the value of the completed investment.

Earlier reference was made under the heading 'The Quantity Surveyor', to the valuation of the site or the profit which the developer was expecting to make. A simple appraisal has been set out where the costs are known and the site is sold to the developer at £500,000. Each point is itemized and the purpose of the exercise is to establish the 'surplus' left over which represents the developer's profit. In this particular development he is aiming for a 20% return on his total costs.

GENERAL DATA

PROJECT APPRAISAL

PROJECT DETAILS

Scheme: Manchester
Client: Roker Developments
Tenure: Freehold
Date: 7/4/90
Reference: BOH

INVESTMENT VALUE	£	£
1. Industrial Warehouse – 60,000 sq.ft		
2. Rent – £5 per sq.ft		
Total rent 60,000 x £5	300,000	
3. Less ground rent	–	
Net estimated rental value	300,000	
4. Year purchase – in perpetuity @ 8.5% =	11.765	
Gross value (£300,000 x 11.7647)	3,529,410	
5. Purchaser's costs – 2.75%	94,461	
Net investment value	3,434,951	

EXPENDITURE

Site purchase

6. Site cost			500,000	
7. Acquisition fees	2.75%		13,750	
8. VAT on both above costs	15%		77,063	

Pre-development costs

9. Site survey		5,000	
10. Planning fees		2,500	
11. Demolition		10,000	
VAT on above costs		2,625	610,938

Construction costs

12. Estimated building costs 60,000 sq.ft. @ £27.50		1,650,000	
13. Design fees	12%	198,000	
14. Contingency	3%	55,440	
15. Building regulation fee		4,000	
16. Fund supervision fee		–	
VAT on above costs	15%	286,116	2,193,556

                                    Carried forward   £2,804,494

THE DEVELOPMENT PROCESS

| | | | Brought forward | | £2,804,494 |

Marketing costs

17. Letting percentage of estimated rental value	15%	45,000	
18. Brochure and marketing		15,000	
19. Agent's sale fees % of net investment value	1.5%	51,524	
20. Legal sale fee % of net investment value	0.5%	17,175	
VAT on above costs @ 15%		19,305	148,004

Finance

21. Finance rate @ 16%		
22. Site and pre-development costs 100% for 15 months	132,361	
23. Construction and design costs 50% for 9 months	132,921	265,282

Void Finance

24. On estimated rental value for 3 months	75,000

VAT

25. VAT to be calculated on all costs @ 15%	
Total development costs	3,292,780
26. VAT refund	385,108
Total development costs (net of VAT)	£2,907,672
27. Surplus (£3,434,951 - 2,907,672)	£527,279

Summary

Net Investment Value	£3,434,951
Site Value	£500,000
Surplus	£527,279
Surplus on total costs	18.13%
Surplus as months rent	21.09 months
Yield on cost	10.32%

N.B. Some of the above figures may have suffered from slight distortion due to rounding off.

Item 1. The gross building area is 60,000 sq.ft.

Item 2. £5 per sq.ft. represents the estimated rental value.

Item 3. The site is freehold and therefore no ground rent needs to be deducted.

Item 4. The agent and his developer believe that 8.5% yield is appropriate for this type of development.

Item 5. The investor who ultimately acquires the development will have his own costs, i.e. agents, solicitors, stamp duty, and will therefore require to deduct these off the gross development value to provide him with his net investment value.

GENERAL DATA

Item 6. Site Cost - £500,000 is the figure at which the site can be bought in the open market.

Item 7. The developer will have his own acquisition costs being agents, solicitors and stamp duty.

Item 8. VAT is charged on all costs in accordance with recent legislation but is recovered by the developer throughout the scheme.

Item 9. It is likely that the developer will need a survey of the site.

Item 10. The scheme is fairly straight forward and no costs have been allowed for a planning appeal. Therefore the planning fees are reasonably nominal.

Item 11. There is a solid structure on the site and therefore an allocation of £10,000 has been made for demolition, clearance and grubbing out of foundations.

Item 12. At competitive tender the developer and his quantity surveyor believe that £27.50 per sq.ft. is a fair building price. No differential is made between the 60,000 sq.ft. in the investment value and the 60,000 sq.ft. in terms of building costs. If this was an office building then one would allow of the order of 85% from a gross to net area.

Item 13. Design fees are made up of architects, quantity surveyors and engineers and would be of the order of 12% of the estimated building cost.

Item 14. Contingency - an allowance has been made of 3% for contingencies, that is any unknowns which are found during construction.

Item 15. Over and above planning permission, building regulations are required and an allocation of £4,000 has been made.

Item 16. As explained earlier in the chapter the development has now been forward funded and therefore the fund has no supervision fee in this particular scheme.

Item 17. The developer has decided to bring in a joint agent. For sole agency the fee would be 10% of 1 years full rental value. However in this case with a joint agent the fee will be 15% and the developer hopes that by having 2 firms letting the units he will secure good quality tenants much sooner.

Item 18. The joint agents will have a budget of £15,000 for brochure, boards and advertising.

Item 19. When the development is completed and let, the agent will sell the development to an investor and his fee will be 1.5% of the net investment value.

Item 20. A solicitor will be required to draw up the contract on behalf of the developer when selling to the investor and his fees will be of the order of .5%.

Item 21. The developer is able to secure interest at 2% above base rate.

THE DEVELOPMENT PROCESS

Item 22.   The developer must pay the full rate of interest on the site and his costs. 15 months has been taken which is made up of say 3 months between owning the site and obtaining planning permission, building regulations and commencing construction, a construction period of 9 months to complete the development, and a further 3 months post-completion for letting and selling the investment.

Item 23.   During the 9 month construction period the costs of building together with all the design costs will attract interest on a cash flow basis. In other words, not all of the money is outstanding for the entire period but part of the money is outstanding part of the time. A simplistic view has been taken that 50% of the total monies will be outstanding for the whole 9 months building period.

Item 24.   3 months has been allowed for void finance. This is a calculated decision in that the development is in a prime part of Manchester and should let very quickly but an allowance must be made either for a void period or by way of offering a rent free incentive to an incoming tenant.

Item 25.   VAT. As mentioned earlier in the chapter VAT is levied in accordance with the regulations but is fully recovered during and after the development period.

Item 26.   As above.

Item 27.   The surplus or "what is left for profit" represents 18.13% of total costs. This is slightly less than the 20% which the developer requires but on reflection he is prepared to go ahead in order to secure this prime development opportunity.

14 Professional fees

ARCHITECTS' FEES

The following is a precis of Part 4 of the Conditions of Appointment published by the Royal Institute of British Architects which contains graphs of the fees recommended but not mandatory for professional services. There are two basic categories.

1. New work
2. Work to existing buildings

In each category there are 5 classes of buildings ranging in complexity from simple storage sheds to hospital laboratories. Fees are graded accordingly and recommended percentages shown in the form of graphs. From the graphs the folowing examples have been derived as an indication of the variations between value of work and classes of work.

New Work

Value of Contract Work £	Class 1 Simple sheds %	Class 2 Speculative factories and offices %	Class 3 Purpose built premises %	Class 4 Civic Centres, Hospitals %	Class 5 Courts, Laboratories %
50,000	7.20	7.75	8.25	9.00	9.75
100,000	6.50	7.00	7.55	8.20	8.80
150,000	6.20	6.70	7.25	7.80	8.45
250,000	5.85	6.30	6.80	7.45	8.00
750,000	5.30	5.80	6.25	6.80	7.30
1,000,000	5.25	5.70	6.20	6.70	7.25
2,000,000	5.15	5.60	6.05	6.55	7.05
5,000,000	5.05	5.55	6.00	6.50	7.00

GENERAL DATA

Work to existing buildings

Value of Contract Work £	Class 1 Simple sheds %	Class 2 Speculative factories and offices %	Class 3 Purpose built premises %	Class 4 Civic Centres, Hospitals %	Class 5 Courts, Laboratories %
50,000	10.55	11.50	12.30	13.20	14.10
100,000	9.75	10.55	11.40	12.25	13.15
150,000	9.30	10.10	10.90	11.75	12.60
250,000	8.85	9.65	10.35	11.25	12.00
750,000	8.20	8.85	9.60	10.35	11.15
1,000,000	8.10	8.75	9.45	10.25	11.00
2,000,000	7.80	8.50	9.15	9.90	10.70
5,000,000	7.75	8.35	9.00	9.75	10.50

For works less than £20,000 and over £5 million the architect should agree the fee basis at the time of the appointment.

In addition to the above percentage fees, all expenses and disbursements for printing, models, photographs, travelling expenses, cost of postage, telephone calls and fax messages will be reimbursed where properly incurred in connection with the appointment. Alternatively a percentage addition or lump sum may be agreed.

PROFESSIONAL FEES

QUANTITY SURVEYORS' FEES

The following worked examples of fees are based upon information contained in Professional Charges for Quantity Surveying Services price £5.45 (including postage) obtainable from Surveyors Publications, Norden House, Basing View, Basingstoke, Hants. RG21 2HN (0256 55234). It should be noted that these scales are recommended not mandatory.

Scales 36 and 37 - Building Work

There are three basic categories of works

        Category A. Complex with little repetition

        Category B. Less complex with some repetition

        Category C. Simple

and two basic scales of fees

    1. Inclusive scale for complete service - Scale 36

    2. Itemized scale divided into pre- and post-contract services - Scale 37

Worked examples of the percentages and actual fees are shown below.

Scale 36   Inclusive services

Value of Work £	Cat. A £	Cat. A %	Cat. B £	Cat. B %	Cat. C £	Cat. C %
150,000	9,380	6.25	9,060	6.04	7,650	5.10
250,000	14,380	5.75	13,760	5.50	11,750	4.70
350,000	19,030	5.44	18,060	5.16	15,450	4.41
450,000	23,330	5.18	21,960	4.88	18,750	4.17
750,000	34,880	4.65	32,010	4.27	27,450	3.66
1,250,000	51,880	4.15	46,010	3.68	39,950	3.20
2,500,000	90,380	3.62	79,010	3.16	68,200	2.73
4,000,000	133,380	3.33	116,010	2.90	99,200	2.48

GENERAL DATA

Scale 37   Pre-contract services

Value of Work £	Cat. A £	Cat. A %	Cat. B £	Cat. B %	Cat. C £	Cat. C %
150,000	4,730	3.15	4,410	2.94	3,930	2.62
250,000	7,030	2.81	6,410	2.56	5,730	2.29
350,000	9,080	2.59	8,160	2.33	7,230	2.07
450,000	10,880	2.42	9,660	2.15	8,430	1.87
750,000	15,830	2.11	13,560	1.81	11,580	1.54
1,250,000	23,330	1.87	19,060	1.52	16,080	1.29
2,500,000	39,080	1.56	31,810	1.27	26,330	1.05
4,000,000	56,080	1.40	45,810	1.15	37,330	0.93

Scale 37   Post-contract services (overall charges - Alternative 1)

Value of Work £	Cat. A £	Cat. A %	Cat. B £	Cat. B %	Cat. C £	Cat. C %
150,000	3,150	2.10	3,150	2.10	2,520	1.68
250,000	4,850	1.94	4,850	1.94	4,020	1.61
350,000	6,500	1.86	6,450	1.84	5,470	1.56
450,000	8,100	1.80	7,950	1.77	6,870	1.53
750,000	12,450	1.66	11,850	1.58	10,620	1.42
1,250,000	18,950	1.52	17,350	1.39	16,120	1.29
2,500,000	34,200	1.37	30,100	1.20	27,870	1.11
4,000,000	51,200	1.28	44,100	1.10	40,370	1.01

For negotiating and agreeing prices with a contractor	Value of work £	Fee £	%
	150,000	750	0.50
	250,000	1,050	0.42
	350,000	1,350	0.39
	450,000	1,650	0.37
	750,000	2,400	0.32
	1,250,000	3,350	0.27
	2,500,000	4,600	0.18
	4,000,000	6,100	0.15

Scale 38 Civil Engineering Works

Category I – Runways, roads, railways and earthworks and dredging and monolithic walls.

Category II – Piled quay walls, suspended jetties, bridges, sewers, storage and treatment tanks, turbine halls, reactor blocks.

Pre-contract services

Value of Work £	Cat. I Fee £	%	Cat. II Fee £	%
500,000	1,960	0.65	3,650	0.73
750,000	2,790	0.37	5,100	0.68
1,500,000	5,040	0.34	8,850	0.59
2,500,000	7,540	0.30	12,850	0.51
5,000,000	12,790	0.26	21,850	0.44
7,000,000	16,790	0.24	28,850	0.41
12,000,000	25,790	0.21	45,350	0.38
15,000,000	30,290	0.20	54,350	0.36
25,000,000	44,790	0.18	83,350	0.33

Post-contract services

Value of Work £	Cat. I Fee £	%	Cat. II Fee £	%
500,000	5,950	1.19	10,750	2.15
750,000	8,250	1.10	15,000	2.00
1,500,000	14,250	0.95	26,250	1.75
2,500,000	20,750	0.83	38,250	1.53
5,000,000	34,000	0.68	65,250	1.31
7,000,000	44,000	0.63	86,250	1.23
12,000,000	68,000	0.57	135,750	1.13
15,000,000	81,500	0.54	162,750	1.09
25,000,000	125,000	0.50	249,750	1.00

GENERAL DATA

CONSULTING ENGINEERS' FEES

Agreement 2 for Civil, Mechanical and Electrical Work and for Structural Engineering Work where an Architect is not appointed by the client.

The following is a precis of the payment for normal services described in clause 10.1 of the ACE Conditions of Engagement 1981 published by the Association of Consulting Engineers. It should be noted that the scales are recommended and not mandatory.

10.1 Payment depending upon the actual cost of the Works.

This method of payment is recommended for use in normal circumstances.

10.1.1 The sum payable by the Client to the Consulting Engineer for his services under Clause 6 shall be calculated as follows:-

(a) The Works shall first be classified into one or more of the following classes as shall be appropriate:

Class A Civil and structural engineering including Class D work incidental thereto and geotechnical investigations.

Class B Buildings including Class D work incidental thereto and building and engineering services associated with buildings.

Class C Mechanical and electrical plant and equipment.

Class D Structural work in reinforced concrete, prestressed concrete, steel and other metals.

(b) The cost of each class of work shall next be calculated and

(c) The sum payable by the Client to the Consulting Engineers shall then be determined and shall be an amount or the sum of the amounts calculated in respect of each relevant class of work in accordance with the percentages stated in Article 4(a) of the Memorandum of Agreement.

Agreement 3 for structural engineering work where an Architect is appointed by the client.

The following is a precis of the payment for normal services described in clause 10 of the ACE Conditions of Engagement 1981 published by the Association of Consulting Engineers. It should be noted that the scales are recommended and not mandatory.

10.1.1 Payment depending upon the actual cost of works

The fee calculation is divided into three main parts:-

a) Cost of the works including reinforced concrete and reinforced brickwork or blockwork, excluding unreinforced brickwork or blockwork. This fee is calculated by dividing the cost of the works by an index - Output Price Index published by the Department of the Environment. By applying the resulting figure to a graph a percentage fee is obtained.

b) Cost of reinforced concrete work and/or reinforced brickwork or blockwork. The fee for reinforced concrete and brickwork is usually 3% of the value of those elements.

PROFESSIONAL FEES

c) Cost of the unreinforced load bearing brickwork or blockwork. The fee is usually 3.5% of the value of this element.

In addition to normal fees other payments can include for the use of computer or special equipment, site supervision of staff resident on site, disbursements including extra professional indemnity insurance taken out in accordance with wishes of the client.

CONSULTING ENGINEERS' TYPICAL FEES

Example of fees (Agreement 2)

		£
Class A	Value of civil and structural engineering including Class D work	5,956,250
Class B	Value of buildings including Class D work	1,125,500
Class C	Work excluded	–
Class D	Value of structural work	4,856,500

Calculation:-

Class A $\dfrac{5,956,250}{321}$ = (18,555)   5.5%

Class B $\dfrac{1,125,500}{321}$ = (3,506)   6.75%

Class D $\dfrac{4,856,500}{321}$ = (15,129)   2.25%

	£		£
Total fee	5,956,250 @ 5.50% =		327,594
	1,125,500 @ 6.75% =		75,971
	4,856,500 @ 2.25% =		109,271
	Represents an average of 7%		£ 496,577

Example of fees (Agreement 3)

Clause   10.1 Payment depending upon the actual cost of the work

10.1.1. (a) (1) $\dfrac{6,000,000}{321}$ = 18,692 say 18,700 = 5%

(2) $\dfrac{750,000}{321}$ = 2336 say 2300 = 6%

(3) $\dfrac{500,000}{321}$ = 1558 say 1560 = 6.30%

(4) $\dfrac{250,000}{321}$ = 779 say 780 = 6.90%

GENERAL DATA

plus

10.1.1. (b)   Normally   3% of RC value and reinforced brickwork

plus

10.1.1. (c)   Normally   3.5% of unreinforced brickwork

Example of fees (Agreement 3)   £

1. Factory units   value   £4,737,952 divided by 321 = 5.1%   241,635

   Reinforced concrete   £1,257,465 @ 3%   37,724

   Brickwork   £210,150 @ 3.5%   7,355

                                                                   £   286,714

                                 Average = 6.05%

                                                                            £
2. Pumping station   value   £2,456,950 divided by 321 = 5.40%   132,675

   Reinforced concrete value   £1,756,450 @ 3%   52,694

   Brickwork   £10,420 @ 3.5%   365

                                                                   £   185,734

                                 Average = 7.56%

The Output Price Index has been taken as 321 for 1st quarter 1990. The latest figure can be obtained from the Association of Consulting Engineers, Alliance House, 12 Caxton Street, London SW1H 0QL.

PROFESSIONAL FEES

LANDSCAPE CONSULTANTS' FEES

Fees on the percentage basis

The following is a precis of the Landscape Institute Conditions of Engagement and Professional Charges. It should be noted that the fees are recommended and not mandatory.

Remuneration on the percentage basis (contracts over £10,000)

The fee is in two parts

Part 1 is assessed from a graph which indicates the fee percentage from 6% to 14% varying according to the value of the contract from £10,000 to in excess of £900,000.

Part 2 is a coefficient ranging from 1.0 where the consultant has overall responsibility to his client for a normal balance of 'hard' and 'soft' works. This may increase to 1.2 when the 'soft' works element exceeds 50% of the landscape contract or for private garden contracts. The coefficient may be decreased to 0.8 for other types of jobs such as golf courses and road landscaping.

Example

Assume a project which has a contract value of £100,000 including both 'hard' and 'soft' works in a new business park.

    The fee graph shows that the percentage 'norm' is 7.5%
Coefficient for soft works element exceeds 50% = 1.2
Job coefficient is 1.0
Compounded coefficient 1.0 x 1.2
Total percentage fee is 7.5% x 1.2 = 9.0%
Fee to be charged is 9.0% of £100,000 = £9000

    The scale of fees allows for other methods of remuneration such as lump sum fees, using a ceiling figure in conjunction with a time basis or having a retainer which can be reviewed after a period and paid according to actual work carried out or allowed to stand in full.
    Similarly when only occasional work is required this can be carried out on a time basis.
    If Bills of Quantities are required these would normally be charged at RICS scale Category C. Site surveys would normally be paid on a lump sum basis of estimated time involved. Disbursements would of course be charged at cost and include normal out-of-pocket expenses.

GENERAL DATA

Professional Team 'All-in' Fee

Assessing the fees for a professional team working on a development project can be complicated due to the different methods of fee calculation adopted by the various professional bodies. Worked examples of these are shown in this chapter related to a range of contract values. Sometimes one discipline is appointed 'lead professional' and an overall fee is agreed with the client for the whole team and the table below shows the effect of this arrangement. The figures have been rounded off to the nearest £1000.

Project Cost £000	8% £000	9% £000	10% £000	11% £000	12% £000	13% £000	14% £000	15% £000
200	16	18	20	22	24	26	28	30
300	24	27	30	33	36	39	42	45
400	32	36	40	44	48	52	56	60
500	40	45	50	55	60	65	70	75
600	48	54	60	66	72	78	84	90
700	56	63	70	77	84	91	98	105
800	64	72	80	88	96	104	112	120
900	72	81	90	99	108	117	126	135
1,000	80	90	100	110	120	130	140	150
1,200	96	108	120	132	144	156	168	180
1,400	112	126	140	154	168	182	196	210
1,600	128	144	160	176	192	208	224	240
1,800	144	162	180	198	216	234	252	270
2,000	160	180	200	220	240	260	280	300
2,250	180	202	225	247	270	292	315	337
2,500	200	225	250	275	300	325	350	375
2,750	220	247	275	302	330	357	385	412
3,000	240	270	300	330	360	390	420	450
3,250	260	292	325	357	390	422	455	487
3,500	280	315	350	385	420	455	490	525
3,750	300	337	375	412	450	487	525	562
4,000	320	360	400	440	480	520	560	600
4,500	360	405	450	495	540	585	630	675
5,500	440	495	550	605	660	715	770	825
6,000	480	540	600	660	720	780	840	900
7,000	560	630	700	770	840	910	980	1050
8,000	640	720	800	880	960	1040	1120	1200
9,000	720	810	900	990	1080	1170	1260	1350
10,000	800	900	1000	1100	1200	1300	1400	1500

15

Construction indices

Indices have been used for many years in the construction industry, not only to assist in obtaining up to date estimates of cost but also in the 'formula method' of calculating the increase or decrease in costs on a contract.

The indices are also useful when projecting cash flow forecasts on a contract which may run over several years.

To use an index, one must have knowledge of the various indices published and select the most appropriate for the type of building or civil engineering work in the contract. A complete guide and set of data is given in Spon's Construction Cost and Price Indices Handbook, M C Fleming and B A Tysoe. (obtainable from E & F N Spon, 2-6 Boundary Row, London SE1 8HN).

Care must be taken when using an index to update rates in a bill of quantities for the purpose of pricing an estimate, that the level of preliminaries relates precisely. If rates include preliminaries in the contract bills it is vital that the index applied also includes them.

By using the elemental cost analyses contained in Chapter 2 and selecting an appropriate index for the year/month/quarter a cost plan can be prepared which then could be applied to a new building based on the rate per square metre of each element.

The main source of indices for the quantity surveyor is the Building Cost Information Service (BCIS) which publishes a range of indices. Other sources include the NEDO indices for the variation of price clauses prepared by the PSA and published by HMSO. (See page 203 for full list). Other important information on building costs and levels is stated in Spon's Architects' and Builders' Price Book (obtainable from E & F N Spon - see above).

The Civil Engineering industry under the ICE Conditions of Contract use a price adjustment formula (sometimes referred to as the Baxter formula) on jobs with a long construction period. The information is published along with the indices for building referred to above on fewer items of materials than the Building Cost indices and include the cost of:

1. Labour and supervision in civil engineering construction
2. Plant - cost of providing and maintaining constructional plant and equipment
3. Aggregates
4. Bricks and clay products
5. Cement
6. Cast iron pipes and fittings
7. Coated roadstone and bitumen products
8. Timber
9. Fuel for plant
10. Reinforcing steel and metal sections
11. Structural steelwork

GENERAL DATA

Construction costs are amounts paid by contractors for the labour employed, materials purchased, plant costs, rates, rent overheads and taxes.

Construction cost indices are compiled by various authorities, institutions and quantity surveying practices; they are published periodically in journals and publications from HMSO. The figures are based on information about changes in material costs, wage rates, etc, to measure changes in costs without any allowance for market trends.

Tender level rates are the amounts charged to clients which fluctuate according to market trends. They increase when there is a large volume of work available and decrease during lean periods. The tender level rates are calculated directly from information in building tenders. It has been known for contractors to put tender prices of nett cost without any addition for profit (and even less than nett) when a period of depression in the industry prevails, in order to keep the work force and plant occupied. Tender level indices are compiled from rates in accepted tenders during a certain period, compared to a standard schedule, and averaged.

The following table shows a comparison between construction costs and tender levels during the period 1970 to 1989.

INDICES

YEAR	CONSTRUCTION COSTS	TENDER LEVEL RATES
1970	100	100
1971	109	115
1972	119	145
1973	140	199
1974	166	237
1975	205	242
1976	241	241
1977	275	258
1978	299	299
1979	342	371
1980	410	458
1981	458	467
1982	506	453
1983	537	469
1984	569	495
1985	600	514
1986	631	540
1987	665	608
1988	706	728
1989	759	851

From these figures it can be seen for example the depression in the industry in 1976 to 1978 and again between 1982 to 1987. (These figures are based upon information contained in Spon's Architects' and Builders' Price Book and Spon's Construction Cost and Price Indices Handbook published by E.and F.N. Spon, 2-6 Boundary Row, London SE1 8HN).

CONSTRUCTION INDICES

It should be mentioned here that PSA have other schedules of rates which are used for Measured Term Contracts and are updated each month by percentage adjustments, not indices. The statistical use of these may be of great use in updating items on maintenance contracts.

Other bodies such as British Telecom use the National Building Schedule which is independently produced.

The use of indices is simply explained as follows:-

1. Take the index of the current year/month/quarter from the relevant table e.g. 1990 1st quarter - 367.

2. Take the index of the year/month/quarter of the tender being updated e.g. 1987 3rd quarter - 259.

$$367 - 259 = 108$$

$$\frac{108}{259} \times 100 = \underline{41.70\%}$$

Rates to be increased by 41.70% for new project but adjustments for regional cost differences must also be considered. (See PAGE xiii).

Indices used in construction industry

Generally	BCIS Royal Institution of Chartered Surveyors 85/87, Clarence Street Kingston Upon Thames, Surrey KT1 IRB (081-546 7554)
DOE Public Sector Building Tender Price Index Published in <u>Housing and Construction Statistics</u>	HMSO
BCIS Building Tender Price Index BCIS General Building Cost Index	Davis, Langdon and Everest, Tender Price Index, published in Spon's Price Books.
PSA (Department of Environment) price adjustment formulae for construction contracts, monthly bulletin of indices	HMSO

It should be noted that other indices relevant to the construction industry are printed in the following books published by E. and F. N. Spon:

Spon's Architects' and Builders' Price Book
Spon's Civil Engineering Price Book
Spon's Landscape and External Works Price Book
Spon's Mechanical and Electrical Services Price Book

16

Property insurance

Rebuilding Costs for Insurance

It is essential that property owners have adequate insurance cover their property and this section is intended to show how the sum to be covered can be assessed quickly. The easiest way to achieve this is by using the square metre prices in Chapter 1 as a base.

The appropriate rate should be multiplied by the area of the building to be insured. The resultant figure must then be adjusted by applying both the historical indices on page 203 and the regional variation factor on page xii.

The insurance cover must also include for the demolition of the damaged building (not just clearing away debris but grubbing existing foundations and basements) and professional fees to plan and supervise the work of reconstruction.

Example

Pre-war office block, original cost unknown; total loss

		£
Present day cost:		
Office block – 4,000 square metres @ £900m2		3,600,000
Add 5% for geographical variation		180,000
Allowance for period for redesign, obtaining permisions, tendering etc.		3,780,000
2 years @ 7½% p.a.		588,263
		4,368,263
Assuming 18 month building contract with predicted inflation @ 8% p.a. (9 months)		262,096
		4,630,359
Allowance for demolition of old building		200,000
		4,830,359
Allowance for professional fees	15%	724,554
Carried forward	£	5,554,913

205

GENERAL DATA

	Brought forward	£	5,554,913
Add VAT		15%	833,237
Required insurance cover		£	6,388,150

If the original building costs were known (i.e. the costs would be available if the property was constructed recently) the present day costs would be calculated by applying indices to the original building cost figures to bring them up to date and using the cost per square metre method as above.

Index

Administration buildings, 121
Agricultural work, 124, 126
Air conditioning, 124–5, 137
Airport hangers, 5
Alteration work, 171–3
Aluminium
　partitions, 35
　windows, 35
Ambulance stations, 4, 120, 122
Animal breeding units, 120, 122
Artex, 39
Art galleries, 120, 122
Ashlar masonry, 84
Asphalt paving, 155–6

Ballast, 79
Barns, 120, 122
Basecourse, 77
Bath, 40
Beam, 71
Bench seats, 164
Bicycle stands, 164
Bitumen macadam, 77
Blocks lightweight, 35
Blockwork
　partitions, 35
　walls, 84
Boilers
　gas, 134
　oil, 135
Bollards, 164
Bolts, 71
Bonds, 47
Boreholes, 51
Bricks
　common, 30, 83
　engineering, 30, 83
　facing, 30, 83–4

Brickwork
　manholes, 42
　partitions, 35
　paving, 155
　walls, 30, 83
Builders work, 21
Buffer stops, 80
Bullhead rails, 79–80
Bungalows, 5
Bus stations, 5, 120, 122

Cables
　generally, 143
　tray and ladder, 144
　trenches, 69
Calorifiers, 139–40
Carpet tiles, 39
Carriageway, 77
Car showrooms, 5, 7, 26
Cast iron pipes, 42
Ceiling finishes
　artex, 39
　generally, 8–26
　plaster, 39
　skim, 39
　suspended, 40
Ceramic tiles, 39
Chainlink fencing, 87
Chalk, 57
Chipboard, 38
Churches, 120, 122
Cinemas, 4
Civic offices, 14
Clay, 57
Clay tiles, 39
Clinics, 4, 121, 123
Coach screws, 74
Cold water services, 8–26, 40
Colleges, 4

Index

Columns, 71
Communications, 126
Community centres, 4, 120, 122
Computer installations, 120, 122, 124
Compressed air, 124
Concert halls, 120, 122
Concrete
   bases, 27
   beams, 59
   beds, 64
   blinding, 59
   blocks, 31
   channels, 78
   columns, 59
   copings, 61
   dams, 105
   edgings, 78
   floors, 30
   foundations, 28–9, 59
   frame, 30
   generally, 59–61
   haunches, 65
   kerbs, 78
   manholes, 42
   paving, 43, 156, 164
   piling, 75–6
   pipes, 66–7
   precast, 61
   roads, 156, 164
   shaft lining, 82
   sills, 61
   slabs, 27, 59, 61
   stairs, 36
   surround, 65
   tunnel lining, 74, 81
   walls, 27, 31, 59
   weir blocks, 61
Conduits, 141
County courts, 120, 122
Courts
   county, 120, 122
   crown, 120, 122
   high, 120, 122
   magistrates, 120, 122
Curtain walling, 32

Damp proof course, 28
Damp proofing, 86
Dams
   concrete, 105
   earth, 106, 107

Day nurseries, 7, 15
Decking
   hardwood, 74
   softwood, 74
Demolition, 56, 157, 161, 171
Demountable partitions, 35
Department stores, 120, 122
Development
   costs, 186–7
   profit, 187, 189
Diamond crossings, 80
Diaphragm walls, 53
Disabled accommodation, 121, 123
Doctors' clinics, 4
Doors
   internal, 35
   firecheck, 36
   flush, 36
   hardwood, 36
   panelled, 35, 36
   revolving, 35
   rubber, 35
   security, 35
   softwood, 35
   solid cure, 35
   steel, 35
Dormitory hostels, 121, 123
Drainage
   band, 55
   beds, 63
   cast iron pipes, 41
   concrete pipes, 66–7
   ductile iron pipes, 66
   excavation, 62
   french, 68
   generally, 8–26
   gullies, 43, 68
   haunches, 63
   inspection chambers, 43
   manholes, 42
   sand, 55
   surrounds, 64
   uPVC pipes, 66
   VC pipes, 41, 66
   wick, 55
Dredging, 57
Ductile iron pipes, 66
Ducts
   one way, 69
   two way, 69
   trenches, 69

Ductwork, 133, 136–7

Elderly persons' homes, 7, 24
Electrical installation, 8–26
Electricity, 50
Epoxy flooring, 38
Escalators, 149–50
Excavation
 chalk, 57
 disposal, 57, 154
 foundations, 57
 reduced level, 153, 162
 rock, 57
 topsoil, 57, 153, 158, 163
 tunnels, 81
 turf, 153, 158
External works, 8–26

Factories, 3, 7–10, 120, 122, 124, 126
Fans, 138
Fees
 architects, 191–2
 'all-in', 200
 engineer, 196–8
 landscape architects, 199
 quantity surveyors, 193–5
Felt roofing, 33
Fencing
 chain link, 72, 158–9
 gates, 159
 mild steel, 159
 post and wire, 158
 safety, 72
 timber, 72
Filling material, 58, 154, 158
Filtram, 68–9
Finishes
 ceiling, 8–26, 39–40
 floor, 8–26, 38–9
 wall, 8–26, 37–8
Fire
 alarms, 145
 fighting equipment, 138
 protection, 124–8
 stations, 4, 120, 122
Fish plates, 79
Fittings, 8–22
Flats, 5, 7, 22–3
Flooring
 carpets, 39
 linoleum, 39
 metal, 73
 vinyl, 39
 woodblock, 39
Floors, 8–26, 30
Fluorescent lighting, 167
Flush doors, 35–6
Formwork, 59, 156
Frames, 8–26, 30
French drains, 68

Gabions, 87
Garages, 7, 26, 120, 122
Gas services, 8–26
Gates
 hardwood, 87
 posts, 87
 softwood, 87, 159
Geotechnical work, 52
Glasshouses, 120, 122
Granite setts, 155
Granolithic, 38
Granular filling
 beds, 63
 haunches, 63
 generally, 58, 77
 surrounds, 63
Grass
 block paving, 157
 seeding, 158, 163
Greenheart timber, 74
Ground anchors, 54
Ground investigation, 51
Grout holes, 52
Grouting, 53
Gullies, 43, 68
Gutters, 87

Halls of residence, 121, 123
Handrails, 72
Hardcore, 77
Hardwood
 doors, 35–6
 flooring, 38
 windows, 34
Health centres, 4, 7, 16, 121, 123
Heating installation, 8–26
Heating and ventilation, 124–5
Hedges
 beech, 164
 hawthorn, 164
 privet, 164

High courts, 120, 122
Homes for elderly, 121, 123
Hose reels, 139
Hospitals, 4, 121, 123
Hostels, 121, 123
Hotels, 5, 120, 122
Hot water services, 8–26, 40
Houses, 5, 121, 123
Housing association
 bungalows, 5
 flats, 5, 22–3
 sheltered housing, 5
Hydrants, 139

Indices, 201–3
Industrial buildings, 3
Inspection chambers, 43
Insulation, 132
Insurance, 47
Interest, 188

Joints
 formed, 60
 generally, 78, 156
 open, 60

Laboratories, 121, 123
Ladders, 72
Lagoons, 108
Landings, 72
Landscaping
 hard, 154–7
 soft, 161–4
Land values, 185, 188
Latex flooring, 38
Lavatory basins, 41
Leisure centres, 4, 7, 17
Libraries, 120, 122
Life cycle costing, 177–82
Lifts, 5, 8–26, 149–50
Lighting
 cables, 147
 emergency, 126–8
 generally, 126–8, 148
 fluorescent, 147
 points, 147
Lightning protection, 126–8, 145
Linoleum tiles, 39
Litter bins, 164
Local authority housing, 5

Mains power, 126–8, 148
Magistrates' courts, 120, 122
Manholes
 brick, 42, 67
 concrete, 42, 68
Masonry
 ashlar, 84
 rubble, 84
Metalwork
 flooring, 73
 generally, 72
 handrails, 72
 ladders, 72
 landings, 72
 platforms, 72
 stairways, 72
 walkways, 72
Method related charges, 49
Mild steel stairs, 37
Motels, 5
Multi-storey car park, 5
Museums, 120, 122

Nurseries, 7, 19

Offices, 3, 4, 7, 10, 12–14, 120, 122, 124
Openings, 171
Overflows, 8–26
Overheads, 47

Paint, 33, 37, 39
Painting, 85
Panelled doors, 35
Pannelling, 38
Partitions
 aluminium, 35
 block, 35
 brick, 35
 demountable, 35
 generally, 8–26
 stud, 35
 WC, 36
Pavings
 concrete, 43
 flags, 43
 roads, 77
 slabs, 154–5
Petrol stations, 5
Piling
 bored, 75
 driven, 75

Piling (contd.)
  generally, 29, 75–6
  interlocking, 75–6
  isolated, 75–6
  pre-formed, 75–6
Pipes
  copper, 131
  drains, 66–7
  mild steel, 130–1
  overflow, 8–26
  soil, 8–26
  waste, 8–26
Plants, 164
Plant containers, 164
Plaster, 32, 37, 39
Plasterboard, 37
Platforms, 72
Plywood, 38
Police stations, 4, 120, 122
Polytechnics, 4
Power points, 146
Prisons, 120, 122
Project costs
  earth dams, 106–7
  concrete dams, 105
  generally, 97–115
  intakes, 102
  lagoons, 108
  pump houses, 99, 104
  pumping plant, 103
  reservoirs, 108, 110
  sewage treatment works, 112
  sewerage, 98
  shafts, 101
  tunnels, 101
  water mains, 100
  water towers, 111
  water treatment works, 109
Public address systems, 126–7
Public houses, 4, 120, 122

Quarry tiles, 39

Rail track, 79–80
Reinforcement, 59–60, 78, 156
Reinstatement, 69
Research buildings, 121, 123
Reservoirs, 108, 110
Residential homes, 24–5
Restaurants, 4
Retail warehouses, 4

Revolving doors, 35
Roads, 44, 77, 164
Rock, 57
Roof
  asphalt, 86
  built up, 86
  clay tiles, 33
  cladding, 34
  concrete, 33
  decking, 86
  eaves, 33
  felt, 33
  flat, 33
  generally, 8–26
  hip, 34
  pantiles, 33
  ridge, 34
  slates, 33
  steel decking, 33
  verges, 33
Roughcast, 32
Rubber doors, 35
Rubble wall, 84

Safety fencing, 72
Sand beds, 63
Sanitary fittings
  baths, 40
  drinking fountains, 40
  generally, 8–26
  lavatory basins, 40
  sinks, 40
  showers, 40
  urinals, 40
  WCs, 40
Schools, 4, 7, 20, 121, 123
School halls, 21
Screeds
  cement and sand, 38
  epoxy, 38
  granolithic, 38
  latex, 38
Security, 126–8
Security doors, 35
Services
  cold water, 8–26, 40
  hot water, 8–26, 40
Sewage treatment works, 112
Sewerage, 98
Sewers
  interruptions, 88

Sewers (*contd.*)
   renovation, 88
   stabilization, 88
Shafts, 81
Shaver sockets, 146
Sheds, 120, 122
Shingles, 32–3
Sheltered housing, 5
Shops, 120, 122, 124, 126
Showers, 40
Shrubs, 164
Sinks, 40
Site clearance, 56
Sleepers
   generally, 79
   hardwood, 79
   softwood, 79
Softwood
   doors, 35
   flooring, 38
   windows, 34
Soil pipes, 8–26
Solid core doors, 35
Specified requirements, 48
Spiral stairs, 37
Sports facilities, 4, 7, 17, 121, 123
Squash centres, 121, 123
Stairs
   concrete, 36
   generally, 8–26
   metal, 72
   mild steel, 37
   softwood, 37
   spiral, 37
   stainless steel, 37
Steel
   doors, 35
   windows, 34
Storage heaters, 146
Strip foundations, 28–9
Structural steelwork
   anchorages, 71
   beams, 71
   blast cleaning, 71
   bolts, 71
   bracings, 71
   columns, 71
   galvanizing, 71
   generally, 71
   painting, 71
   permanent erection, 71

   portal frames, 71
   purlins, 71
   trusses, 71
   wire brushing, 71
Substructures, 8–27
Supermarkets, 4, 120, 122
Superstores, 7, 15
Superstructures, 30
Supervision, 50
Surgeries, 121, 123
Suspended ceilings, 40
Swimming pools, 4, 121, 123
Switches, 79
Switchgear, 126–8, 148

Tarmacadam, 155, 164
Telecommunications, 124
Telephones, 126–8
Temporary works, 49
Terrazzo tiles, 39
Theatres, 120, 122
Tiles
   carpet, 39
   ceramic, 38–9
   clay, 33
   concrete, 33
   cork, 39
   hanging, 32
   quarry, 39
   terrazzo, 39
   vinyl, 39
Timber
   decking, 74
   fencing, 86
   generally, 74
   greenheart, 74
Topsoil, 57–8
Training colleges, 4
Transformers, 146
Transport
   facilities, 5, 122
   garages, 120
Trees
   mature, 163
   pits, 163
Trial holes, 51
Trunking
   PVC, 142–3
   steel, 142
Tyrolean finish, 32

Universities, 4, 121, 123
Underpinning, 29
uPVC pipes, 67
Urinals, 40

Ventilation installation, 8–26
Vinyl tiles, 39
VC pipes, 66

Walkways, 72
Walls
  curtain, 32
  external, 8–26, 32
  finishes, 8–26, 37–8
  internal, 8–26, 32
Wallpaper, 33
Warehouses, 3, 4, 7, 11, 120, 122
Waste pipes, 8–26
Water
  cold services, 8–26, 40
  generally, 124–5
  hot services, 8–26, 40
  mains, 100
  plants, 136
  treatment works, 109
  towers, 111
Wearing courses, 77
Waterproofing
  asphalt, 86
  damp proofing, 86
  roofing, 86
  synthaprufe, 86
WC partitions, 35
WC suites, 40
Windows
  aluminium, 35
  double glazed, 34
  generally, 8–26
  hardwood, 34
  softwood, 34
  steel, 34
  uPVC, 35
Wood block flooring, 39

York stone, 164